A Glance of Soil Science Research in Manipur - North East India

THE EDITORS

Dr. Indira Sarangthem (M.Sc., Ph.D.) specializes in the field of soil chemistry/soil fertility research, teaching and extension activities for the last 30 years. She has vast academic experience having guided more than 50 M.Sc. and Ph.D. Students. She has handled 17 external funded and intramural projects and is currently Principal Investigator of All India Coordinated Research Project on Micro and Secondary Nutrients and Pollutant Elements in Soils and Plants). She has published 47 full length research papers in national and international peer-reviewed journals, 21 popular articles for the benefit of farmers and other stakeholders, 18 papers in conference proceedings / book chapter, 7 books- authored or edited individually or jointly, 9 manuals/ teaching aids and 8 research bulletins/ extension bulletins. She has attended/presented research outputs /participated in 60 farmers training seminars for dissemination of proven technologies, delivered lectures in 45 summer courses/trainings as Resource person and 29 Radio talks programmes on All India Radio. As a member of ISSS, she has been associated with the society since her student time. She is working as Dean, College of Agriculture, Central Agricultural University, Imphal, Manipur since August 2019. She also is the Chairperson of Faculty of Agriculture of Central Agricultural University, Imphal. Recently she has been nominated by CAU, Imphal to visit Asian Institute of Technology, Bangkok, Thailand to explore the possibilities of collaboration between CAU Imphal and institutes of South East Asian Countries in the fields of mutual interest like Agroecology, Natural Farming and novel Sustainable farming systems.

Dr. Lokesh Kumar Mishra obtained his Post-Graduate and Doctorate degree in Biochemistry from Narendra Deva University of Agriculture and Technology, Faizabad, Uttar Pradesh. He started his academic career as Assistant Professor in Biochemistry at ND University, Faizabad. He later served as Assistant Professor at VBS Purvanchal University Jaunpur, UP. Presently he is working as Associate Professor (Biochemistry) at College of Agriculture, Central Agricultural University, Imphal, Manipur. He is involved in teaching Biochemistry for the last seventeen years. He has been working in different administrative and academic committees of Central Agricultural University like Board of Studies for Faculty of Agriculture as Member Secretary, PME Cell as Member. The author has also handled research projects funded by different agencies of the Government of India. The focus of these projects is metabolic profiling of the indigenous food products of the North East India and development of new products utilizing the biodiversity of the NE India. He has authored several research papers and review articles in International and National Journals of repute and is associated with many national and international Societies in the field of Plant Biochemistry. The Author was awarded the prestigious DBT Overseas Visiting Associate at Department of Plant Sciences North Dakota State University, Fargo, United States. The focus of the research at North Dakota State University was exploration of phenolic linked antioxidant activities of food systems for health promotion and combating metabolic syndromes like type-2 diabetes, hypertension. In recognition of his work in the field of medicinal plants and agricultural research and development he has been awarded fellowship for the year 2019 by International Society of Noni Science, Chennai.

Dr. N. Surbala Devi is Teaching UG, PG and Ph.D. courses in the discipline of Soil Science for more than 10 years. She has guided 10 P.G. students and published more than 25 research papers in national/international journals. She has completed two internal funded projects related to soil fertility mapping and micronutrient distribution in North East India.

Prof. A. Herojit Singh has 32 years of teaching experience for UG, PG and Ph.D. courses in the discipline of Soil Science. Prof. Singh is working Head Department of Soil Science and Agricultural Chemistry College of agriculture , CAU Imphal He has Guided 13 P.G. and 1 Ph. D. students in the discipline and published more than 70 research papers in national/ international journals.

A Glance of Soil Science Research in Manipur - North East India

— *Editors* —

Indira Sarangthem
Dean
College of Agriculture, Central Agricultural University
Imphal, Manipur

L. K. Mishra
Associate Professor & Head
Department of BPME

N. Surbala Devi
Assistant Professor
Department of Soil Science and Agricultural Chemistry

A. Herojit Singh
Professor & Head
Department of Soil Science and Agricultural Chemistry

2022

Daya Publishing House®
A Division of
Astral International Pvt. Ltd.
New Delhi – 110 002

ISBN: 9789359191348

Published by : **Daya Publishing House®**
A Division of
Astral International Pvt. Ltd.
– ISO 9001:2015 Certified Company –
4736/23, Ansari Road, Darya Ganj,
New Delhi-110 002
Ph. 011-43549197, 23278134
E-mail: info@astralint.com
Website: www.astralint.com

CENTRAL AGRICULTURAL UNIVERSITY

IROISEMBA, IMPHAL-795004, MANIPUR (INDIA)

Dr.Anupam Mishra
Vice-Chancellor

Tel : (0385) 2415933(O)
2410450(R)
Gram : AGRIVARSITY
Fax : 2410450

FOREWORD

Humans have known for centuries that healthy, living soils support resilient farming systems with greater yield stability in the face of unpredictable weather extremes and other stresses. Managing soil health reduces crop production and financial risks, and therefore constitutes good business management as well as efficient environmental maintenance. Soils across the globe are degrading at dangerously fast rate. The situation is even worse considering the fact that it takes a thousand years for just one centimetre of topsoil to form. The situation is similar in the North Eastern region of India which is a flourishing Biodiversity hotspot supporting diverse flora and fauna. Soil science researchers all over the globe and particularly North Eastern region of India must support farmers with novel methods that conserve soil health and support farming practices that promote sustainable use of soil. PG and Ph.D. research conducted by students in Agricultural Universities in India provide an effective platform that can be utilized by Agricultural Scientists and academicians to explore the possibilities of developing novel, efficient research strategies in diverse fields including soil science that address the issues faced by farmers and other stakeholders involved in farming practices. Research that demonstrates the importance of soil, efficient use of fertilizers depending on soil health, and protection of soil must be done and disseminated widely. This shall help in encouraging farmers in adopting these soil friendly practices of crop production recommended by PG and Ph.D. research.

Several books and hundreds of papers have been published on soil science. This publication having focus on soil health of North Eastern Region of India particularly Manipur will make this edited book ***(A Glance of Soil Science Research in Manipur – Northeast India)*** a unique and valuable reference for academicians, policy makers, and other stakeholders in the field of soil science. The research findings included in the book should convince the readers about different efficient ways of managing soil health in different farming systems without damaging the soil ecosystem. I congratulate the editing team led by Professor. Indira Sarangthem (Dean, College of Agriculture, Central Agricultural University, Imphal, Manipur) for preparing this comprehensive and informative compilation. The compilation is a dedicated effort to present the different aspects of the "Soil Science Research" in a very concise and relevant manner. I believe that the students and scientists working in the area of Soil Science will find this book as a source of useful and authentic information.

I appreciate the efforts of editing team in compiling this important document.

Date: 29th, March, 2022

(Anupam Mishra)

Preface

This edited book has been compiled with a novel intention of compiling the ground level research being conducted in the field of Soil Science by research workers and students. This being one of the series has tried to compile the different aspects of Soil Science research being carried out in the North East Region of India and particularly the state of Manipur. The contributors of the book chapter deserve all the praise for sharing their research in a concise manner as book chapter for this book. This compilation shall be of great help to students, research workers and policy makers equally in accessing the latest developments in the field of soil science research in different crops in North Eastern Region of India.

The editorial team wishes to thank the publisher "Astral International Pvt. Ltd." for their constant support and suggestions. Their timely cooperation at each stage has helped in bringing this publication from editor's desktop to book form. The contributors and the editors have taken utmost care to avoid any errors, however any descripency in reporting of facts and mis-interpretation pointed out by readers shall be higly welcome.

Indira Sarangthem
L. K. Mishra
N. Surbala Devi
A. Herojit Singh

Contents

Section C: Nutrient Management in Legumes

Section D: Nutrient Management in Vegetables/Citrus Fruits

Introduction

Soils are one of the most diverse habitats on Earth. Soils constitute the habitat for the organisms that inhabit them for a selected period or throughout their entire life cycle. Soils are a vital life sphere that can generate indispensable resources and goods to supply natural and human ecosystems. The numerous biogeochemical cycles and different interactions among spheres (atmosphere, geosphere, biosphere, anthroposphere, and hydrosphere) all converge into the pedosphere. Healthy soil is the foundation for profitable, productive, and climate resilient agricultural systems. Knowledge and awareness about the plant growth supporting soil processes, which impact environmental quality due to adoption of area specific management practices helps in designing a crop and soil management system that conserves and improves soil health in a sustainable manner. Soil health plays an important role in enhancing crop resilience to biotic and abiotic stresses.

Soil health deterioration is an alarming situation in the NE Regions of India facilitated by severe soil erosion on hills by deforestation and shifting cultivation and flood hazards in Brahmaputra valley regions. The region is largely rain-fed and mono-cropped in most of its parts, responsible for steady decline in soil fertility status. Strong soil acidity in hill regions coupled with depletion of soil nutrient status (mainly nitrogen, phosphorus and zinc) in valleys are the major soil health concerns.

Soil Health Management (SHM) is one of the most important interventions under National Mission for Sustainable Agriculture (NMSA). SHM aims at promoting Integrated Nutrient Management (INM) through judicious use of chemical fertilisers including secondary and micro nutrients in conjunction with organic manures and bio-fertilisers for improving soil health and its productivity; strengthening of soil and fertiliser testing facilities to provide soil test based recommendations to farmers for improving soil fertility; ensuring quality control requirements of fertilisers, bio-fertilisers and organic fertilisers. Soil Health Management

In recent years, climate change, intensification of agricultural activities, urbanization, military conflicts, mining and other industrial activities and associated infrastructures for transport, such as roads and railroads among others, have affected soils over many centuries. However, these have become more and more intense during the last two decades, threatening the few spots that still have natural relatively unaltered soils. Agriculture Universities in India having Soil Science and Agriculture Chemistry Departments have been at the forefront of carrying out basic ground level research in the field of soil science as per the local demand from farmers, ecological conditions and other stakeholders. This book compiles the research done by College of Agriculture, Central Agricultural University, Imphal in the field of soil sciences through PG and Ph.D. programs it has been offering since 2002.

Prof. Indira Sarangthem

Dr. L. K. Mishra

Dr. N Surbala Devi

Dr. A Herojit Singh

Section A

Nutrient Management in Paddy

Chapter 1

Nitrogen Transformation in Paddy Soil Fertilized with Organic Manures and Urea

Khundrakpam Manorama

ABSTRACT

A field experiment was conducted to investigate transformation of different inorganic and organic forms of nitrogen, N uptake and yield of paddy on application of different organic manures (vermicompost, poultry manure and FYM) and urea at different growth stages of rice. The treatments consisted of addition of different organic manures viz., vermicompost, poultry manure and FYM applied at 100 per cent RDN and combination of above organic manures @75 per cent and 50 per cent RDN along with urea @25 per cent and 50 per cent RDN, respectively, besides 100 per cent RDN as urea and control. Results revealed that inorganic N (NH_4^+ and NO_3^-) in soil was higher at the start of the growing season and decreased with crop growth. Higher values of NH_4^+ and NO_3^- in soil were observed in plots, which received both organic manures and urea compared to their individual application. Irrespective of different treatments, different organic forms of N viz., amino acid N, hexosamine N, serine+threonine N increased with crop growth. Whereas, total N and total hydrolysable N including hydrolysable ammonium N and unidentified hydrolysable N gradually decreased with crop growth indicating that these fractions mainly contribute to mineralization. Greater amount of different inorganic and organic forms of nitrogen in soil were recorded in singly or combined treated plots of organic manures and urea over control. Organic manures amended soil also resulted in higher build-up of different fractions of N over urea alone. The different forms of N were more under vermicompost-amended soil followed by poultry manure and FYM. Again, irrespective *of different treatments total N uptake reached its peak at 75 DAT and decreased gradually up to the harvest of crop. Comparatively higher N uptake was recorded under different treatment combinations of organic manures and urea over control at all the studied stages. Results further revealed that the addition of organic manures and urea either singly or in combination significantly improved rice yield over control. Moreover, rice yield was more in integrated treatments over their individual addition. Maximum grain yield (5.55 t ha^{-1}) and straw yield (7.23 t ha^{-1}) were recorded in treatment T10 (50 per cent RDN vermicompost + 50 per cent RDN urea) which was on par with T11 (50 per cent RDN poultry manure + 50 per cent RDN urea) and T2 (100 per cent RDN urea) but superior to rest of the treatments. The results indicate that the application of urea-N in combination with organic manures could be considered more effective in rice production for reducing N-losses, conserving-N and increasing the efficiency of applied nitrogen.*

Introduction

Nitrogen (N) is one of the most important plant nutrients required by plants for growth and metabolic activities. It is essential for crop production as it is an important component of chlorophyll, enzymes, amino acids and proteins. Its importance in plant cell, flowering, fruiting and root development and quality improvement is no less than phosphorus and potassium.

Organic and Inorganic Nitrogen are the two forms of nitrogen that exists in soil. A bulk of total nitrogen is present in the organic form and only about 2 per cent in inorganic form. The inorganic form is liable to be lost through different types of losses like run-off, ammonia volatilization, leaching, denitrification and fixation by clay minerals. The organic form of nitrogen, which is hydrolysable is mineralised gradually and is transformed to mineral nitrogen. Nitrogen is transformed in soil tremendously as it is used, re-used and made available by soil microbes. Transformation of nitrogen is a complex process brought about by succession of different microorganisms in the soil which affect the soil fertility. During the transformation, nitrogen is converted from one form to another.

Nitrogen, present in soil, mostly comes from natural sources and applied fertilizers. However, both the sources are important for growing crops. Use of organic inputs is fundamentally important in that they supply various kinds of plant nutrients including micronutrients, improve soil physical and chemical properties which has positive impact on nutrient holding and buffering capacity, and consequently enhance microbial activities. In addition, organic matter continuously releases nitrogen, as plants need it. Estimation of nitrogen mineralization from the manure during a cropping season is important for the adjustment of the requirement of chemical fertilizer to be applied in combination with the manure.

Nitrogen transformations in submerged soils are markedly different from those in drained, aerated soils. These differences affect the prevalent soil microorganisms and microbial activities and the turnover, availability and losses of nitrogen. Anaerobic environment of wet soils results in the accumulation of ammonium, instability of nitrate, and decreased nitrogen requirement for organic matter decomposition (Savant and DeDutta, 1982). Mineralized nitrogen or ammonium released under anaerobic incubation is significantly correlated with soil organic matter; both quality and quantity of organic matter clearly affect nitrogen mineralization in wetland rice soils (Sahrawat, 2006).

Nitrogen is the most important inorganic element impacting the growth and yield of crop. This is especially the case for rice, since 60 to 80 per cent of N absorbed by the plant is derived from the soil N pool (Savant and De Dutta, 1982). The greater part of nitrogen in paddy soil exists in soil organic matter. The mineralization of organic nitrogen or ammonium production is a key process for the nitrogen nutrition of wetland rice (Sahrawat, 2006). The N mineralization rate determines the availability of Nitrogen for the plant growth. An understanding of the behaviour of nitrogen in wetland rice soils is a necessary pre requisite for improving growth and yield and improving the productive efficiency of fertilizer N (Masayna *et al.,* 1985).

Keeping the above information in view, the present research programme entitled "Nitrogen Transformation in Paddy Soil Fertilized with Organic Manures and Urea" were taken up during *Kharif* season of 2015 with the following objectives:

1. To study the changes in different forms of inorganic and organic nitrogen in paddy soil fertilized with organic manures and urea.
2. To study the effect of different organic manures and urea on nitrogen uptake and yield of paddy.

Materials and Methods

Description of Study site

The experiment was conducted at the research farm of the Department of Soil Science and Agricultural Chemistry, College of Agriculture, Central Agricultural University, Imphal, Manipur. The farm is situated at 24048'44.50"N latitude, 93053'29.98"E longitude and at an altitude of 778 m above mean sea level. In general, the climatic condition of Imphal valley falls under sub-tropical. Naturally, the average annual rainfall of this tract is about 1212 mm. Monsoon rain starts from the month of June and continues up to October with pre-monsoon showers from March onward.

Experimental Treatments and Design

a) Treatment Details

There were 11 treatments. The recommended dose of nitrogen (RDN) *i.e.* 60 kg N $ha^{-1}$was met from urea, FYM, vermicompost and poultry manure applied singly or in combination in each treatment except control which received no nitrogen. A constant dose of 40 kg P_2O_5 ha^{-1} through SSP and 30 kg K_2O ha^{-1} through MOP was applied in all the plots except control. The total nitrogen content of the organic manures are FYM- 0.5 per cent, vermicompost- 1.4 per cent and poultry manure- 1.5 per cent. The experiment had following treatments:

T1 -soil (control)

T2 -100 per cent RDN through urea

T3 -100 per cent RDN through FYM

T4 -100 per cent RDN through vermicompost

T5 -100 per cent RDN through poultry manure

T6 -75 per cent RDN through FYM + 25 per cent RDN through urea

T7 -75 per cent RDN through vermicompost + 25 per cent RDN through urea

T8 -75 per cent RDN through poultry manure + 25 per cent RDN through urea

T9 -50 per cent RDN through FYM + 50 per cent RDN through urea

T10 -50 per cent RDN through vermicompost + 50 per cent RDN through urea

T11 -50 per cent RDN through poultry manure + 50 per cent RDN through urea

b) Layout and Design

Experimental design: Randomized Block Design (RBD)

No. of treatments: 11

No. of replications: 3

Total no. of plots: 33

Plot size: 5 m x 4 m

Border between replication: 1 m

Border between plots: 0.3 m

Gross experimental area: 900.37 m^2

Net experimental area: 660 m^2

Rice (*Oryza sativa* L.) cultivar '*Tampha phou*' (CAU-R-1) was taken as the test crop. The cultivar matures in about 140 days. The seeds were sown on May 22, 2015 in seedbeds of about 1.5m in breadth. At the age of 28 days, the seedlings were uprooted and transplanted in the experimental field. The spacing was maintained at 20 cm x 10 cm. The recommended doses of fertilizers *i.e.* 60: 40: 30 N, P_2O_5, K_2O ha^{-1}were applied from different organic sources (FYM, vermicompost and poultry manure) and urea to all the plots except control. Half dose of N, full dose of P_2O_5 and two-third of K_2O in the form of urea, SSP and MOP respectively were applied as basal dressing before transplanting. The remaining half dose of N was applied at tillering and panicle initiation stages equally and the remaining one-third of K_2O was applied at panicle initiation stage. Organic manures *i.e.*, FYM, vermicompost and poultry manure containing equivalent doses of nitrogen were applied 15 days before final land preparation. Standard cultural practices and plant protection measures were taken up as and when needed. The crop was harvested at 115 days after transplanting when about 85 per cent grains reached maturity. It was left in the sun for one week. Then, it was threshed and cleaned by winnowing. Soil samples were collected from 0-15 cm layer at transplanting, 25, 50, 75, 100 DAT (days after transplanting) and at harvest and analysed for exchangeable NH_4^+ and soluble NO_3^-, total N and different organic forms of N. The sequential extraction procedure as given by Bremner (1965) was employed for different organic N fractions including hydrolysable NH_4^+-N, hexosamine-N, serine+threonine-N, amino acid-N, unidentified hydrolysable-N, non-hydrolysable N and total hydrolysable N. Dry matter production was recorded at 25, 50, 75, 100 DAT and at harvest. The plant samples were analysed for N content at 25, 50, 75, 100 DAT and at harvest and corresponding N uptake was recorded by multiplying the N content with dry matter production. Grain and straw yield were recorded at harvest.

Soil Sampling and Data Analysis

The soil sample of the experimental site was collected from 0 to 15 cm depth randomly from several spots of the field before field preparation. A homogeneous composite sample was prepared and analysed for different important physico-chemical properties and nutrient status as mentioned in Table 1.1.

Table 1.1. Physico-chemical Properties of the Soil before Growing the Crop

Soil Property	*Value*	*Methods of Analysis*
Textural class	Clay	Hydrometer method (Jackson, 1973)
Sand (per cent)	9.5	
Silt (per cent)	35.0	
Clay (per cent)	55.5	
pH	5.4	1:2.5 soil:water suspension (Jackson, 1973)
Organic carbon (per cent)	1.26	Walkley and Black titration (Jackson, 1973)
Total Nitrogen (per cent)	0.15	Modified macro - Kjeldahl (Jackson, 1973)
EC (dSm^{-1})	0.22	1:2.5 soil:water suspension (Jackson, 1973)
CEC (cmol(p+) kg^{-1})	14.2	Leaching with 1N NH4OAc (Jackson, 1973)
Available Nitrogen (mg kg^{-1})	141.27	Alkaline potassium permanganate (Jackson, 1973)
Available Phosphorus (mg kg^{-1})	10.52	Bray and Kurtz No.1 (Jackson, 1973)
Available Potassium (mg kg^{-1})	95.03	Flame photometer (Jackson, 1973)
Exchangeable NH_4^+ -N (mg kg^{-1})	47.2	Hesse (1971)
Soluble NO_3^- -N (mg kg^{-1})	20.14	Hesse (1971)
Hydrolysable NH_4^+ (mg kg^{-1})	430	Bremner (1965)
Hexosamine N (mg kg^{-1})	60	Bremner (1965)
Amino acid N (mg kg^{-1})	301.5	Bremner (1965)
Serine + Threonine N (mg kg^{-1})	85	Bremner (1965)

Soil Property	*Value*	*Methods of Analysis*
Unidentified hydrolysable organic N (mg kg^{-1})	440.8	Bremner (1965)
Total hydrolysable organic N (mg kg^{-1})	1232.3	Bremner (1965)
Non-hydrolysable N (mg kg^{-1})	218	Bremner (1965)

Statistical Analysis

All the data pertaining to the present investigation were statistically analysed as per the method described by Gomez and Gomez (1984). The statistical significance of various effects was tested at 5 percent level of probability.

Brief Findings

Nitrogen is typically the main limiting nutrient in the production of lowland rice, and addition of N through organic and inorganic sources is vital for sustained production of rice. The greater part of nitrogen is present in organic form. The organic form of nitrogen, particularly the hydrolysable form, is slowly mineralised and is transformed to mineral nitrogen. Nitrogen mineralization differs for different manure types since the inorganic/organic fraction and quality of organic N varies among manure types. Nitrogen (N) availability from applied manure includes the inorganic N (NO_3^--N and NH_4^+-N) in manure plus the amount of organic N mineralized following application. It is difficult to predict the pattern and amount of nitrogen release from organic nitrogen sources during the growing season. Estimation of nitrogen mineralization from the manure during a cropping season is important for the adjustment of the requirement of chemical fertilizer to be applied in combination with the manure. Present investigation was undertaken to study the changes in different forms of inorganic and organic nitrogen in paddy soil fertilized with organic manures and urea and to study the effect of different organic manures and urea on nitrogen uptake and yield of paddy. The salient findings of the investigation are as follows:

Results indicated that both organic and inorganic forms of N at different growth stages of rice were influenced significantly by the addition of urea either alone or in combination with organic manures. The concentration of exchangeable NH_4^+–N was higher at early stages and progressively declined with advancement of crop growth. The exchangeable NH_4^+–N concentration was higher under combined use of organic manures and urea compared to their individual addition and it was consistently highest with 50 per cent RDN through vermicompost + 50 per cent RDN through urea except at transplanting. Nitrate N in the surface soil at different stages showed lower values in spite of the fact that ammonium N content in soil was relatively high. This might be due to quick loss of nitrate N through denitrification and leaching. Total hydrolysable N decreased gradually with stages of crop growth, which supports the observation that hydrolysable N is more vulnerable to mineralization, and could be considered as a major source of potentially available N for plants than non-hydrolysable N. Among the hydrolysable N fractions, hydrolysable NH_4^+-N and unidentified hydrolysable N gradually decreased with increase in the period of experiment indicating that these fractions mainly contributed to N mineralization. Other fractions *viz.,* amino acid N, serine-threonine N and hexosamine N increased with the period of crop growth. The increase or decrease in different forms of N may be due to interconversion within themselves. Further, results pointed out that the application of organic manures alone or in combination with urea led to a marked increase in different inorganic and organic fractions of N in soil over control. Integrated treatments accumulated greater amount of different inorganic and organic forms of N in soil over organic manures or urea alone. Amongst the integrated treatments T10 (50 per cent RDN through vermicompost + 50 per cent RDN through urea) was found to be superior over other treatments but remained statistically similar with T11 (50 per cent RDN through poultry manure + 50 per cent RDN through urea).

Further study revealed that different treatment combinations of organic manures and urea recorded significantly higher amount of N uptake over control at all the stages. The increased N uptake of different treatments over control was due to increased growth of crop with the application of organic manures and urea. Again, the highest grain (5.55 t ha^{-1}) and straw yield (7.23 t ha^{-1}) were obtained in treatment T10 (50 per cent RDN through vermicompost + 50 per cent RDN through urea) which remained at par with T11 (50 per cent RDN through poultry manure + 50 per cent RDN through urea) and T2 (100 per cent RDN through urea).

Conclusion

Based on the results obtained from the investigation carried out, it could be concluded that within-season inorganic and organic N transformation are governed by the addition of different organic manures and the different levels at which organic manures were applied with urea. From the present study, it is observed that application of different organic sources and urea either singly or in combination resulted in higher accumulation of different inorganic and organic forms of N in soil, higher N uptake and yield over control. Organic manures amended soil also resulted in higher build-up of different fractions of N over urea alone. Integrated treatments performed better than either organic manures or urea alone in respect of grain yield and N uptake. Among the integrated treatments, T10 (50 per cent RDN through vermicompost + 50 per cent RDN through urea) produced highest grain and straw yield which remained at par with T11 (50 per cent RDN through poultry manure + 50 per cent RDN through urea) and T2 (100 per cent RDN through urea). Thus, application of urea-N in combination with organic manures could be considered more effective in rice production for reducing N-losses, conserving-N and increasing the efficiency of applied nitrogen.

REFERENCES

1. Savant, N.K. and DeDatta, S.K. (1982). Nitrogen transformations in wet-land rice soils. *Adv. Agron.,* **35**: 241-295.
2. Sahrawat, K.L. (2006). Organic matter and mineralizable nitrogen relationships in wetland rice soils. *Commun. Soil Sci. Plant Anal.,* **37**: 787-796.
3. Masayna, W., Kai, H. and Kawaguchi, S. (1985). Nitrogen behaviour in tropical wetland rice soil. 2. The efficiency of fertilizer nitrogen, priming effect and A-values. *Fert. Res.,* **6**: 37-47.
4. Bremner, J.M. (1965). Organic forms of nitrogen. In: Black, C.A. *et al.* (eds) Methods of Soil Analysis. *Am. Soc. Agron.,* USA 9, Part **2**, 1238-1253.
5. Jackson, M.L. (1973). Soil chemical analysis. Prentice Hall of India Pvt. Ltd., New Delhi.
6. Hesse, P.R. (1971). A textbook of soil chemical analysis. John Murray (Publishers) Ltd., London.
7. Gomez, K.A. and Gomez, A.A. (1984). Statistical procedures for agricultural research. John Wiley and sons. Inc. London, UK (2nd edtn) pp. 13-175.

Chapter 2

Integrated Nitrogen Management of Rice and Nitrogen Status in Paddy Fields of Imphal West District, Manipur

Takhellambam Sanahanbi Devi

ABSTRACT

An experiment was conducted under laboratory condition to investigate transformation of different inorganic and organic forms of Nitrogen in an acid soil treated with different organic manures [vermicompost (paddy straw and vegetable waste), farmyard manure (FYM) and biogas manure] as nitrogen sources @ equivalent to 50 mg N kg^{-1} soil at different moisture regimes [40 per cent, 60 per cent and 80 per cent water holding capacity (WHC)]. Results revealed that irrespective of different organic manure treatments, significantly higher amount of exchangeable NH_4^+, total N and organic N fractions (unidentified hydrolysable organic N and non-hydrolysable organic N) were found at 40 per cent water holding capacity (WHC). However, at 80 per cent WHC, the magnitude of accumulation of soluble NO_3^--N and other hydrolysable organic N fractions (hydrolysable NH_4^+, hexosamine, amino acid and serine+threonine)-N were comparatively greater than other moisture levels. Results clearly revealed that accumulation of total hydrolysable N was significantly higher at 60 per cent than 40 per cent and 80 per cent WHC. A gradual decease in overtime supports the observation that this fraction is more vulnerable to mineralization and could be considered as a major sources of potentially available N for plants than non-hydrolyzable N. Among the hydrolysable N fractions, unidentified hydrolysable organic N gradually decreased with increase in the *period of experiment indicating that this fraction mainly contributes to N mineralization. Variable trends of different inorganic and organic fractions of N showed that these N forms were affected by moisture regimes. A conversion of one form to another is suggested. Greater amount of different inorganic and organic forms of nitrogen were accumulated in organic matter treated system than untreated one irrespective of moisture regimes and incubation period. Irrespective of different moisture regimes, significantly higher amount of all the inorganic and organic N fractions was accumulated at vermicompost (paddy straw) treated soil. This could be attributed to higher release of inorganic N comparing with other organic N sources resulting in smaller losses of N and building of a higher concentration of total N. The extent of changes of different forms of inorganic and organic N fractions varied with nitrogen fractions, organic matter and moisture regimes. With the proper addition of organic manures at suitable moisture regimes, it may be possible to maintain a constant pool of available nitrogen.*

Introduction

Nitrogen (N) is one of the most important plant nutrient required by plants for growth and metabolic activities. It is essential for crop production as it is an important component of chlorophyll, enzymes, amino acids and proteins. It also encourages vegetative growth. It is required in relatively large concentration by most agricultural crops, but only insufficient quantities are available in the soil. Nitrogen is the most common limiting nutrient in crop production. Its importance in plant cell, flowering, fruiting and root development and quality improvement is not less than phosphorus and potassium.

Nitrogen in soil exists in two major forms: 1) organic and 2) inorganic nitrogen. A bulk of total nitrogen is present in the organic form and only about 2 per cent in inorganic form. The inorganic form is liable to be lost through different types of losses like run off, ammonia volatilization, leaching, denitrification and fixation by clay minerals. The organic forms of nitrogen, particularly the hydrolysable form, is slowly mineralized and is transformed to mineral nitrogen through aminization, ammonification and nitrification processes and becomes available to crops. In contrast, non hydrolysable nitrogen is resistant to mineralization. The stability of some fractions of the organic nitrogen is due to the formation of complex organic molecules, which resist mineralization. Nitrogen undergoes many transformation in soil as it is used, re-used and made available by soil microbes. During the transformation nitrogen is converted from one form to another. In soil, several nitrogen transformation processes may proceed simultaneously and interact in many ways.

Nitrogen, present in soil, mostly comes from natural sources and applied fertilizers. However, both the sources are important for growing crops. But natural source, whether it is from soil organic matter or added organic matters is of utmost importance with respect to maintenance of soil fertility as a whole and controlled release of nitrogen during cropping season in particular. It is critical to understand what happens to organic nitrogen after soil application in order to use it efficiently for crop production. So, understanding the effect of organic amendments on the transformation of nitrogen into different forms is a pre requisite for managing nitrogen input for a given soil.

Nitrogen transformation in soil is influenced by organic matter additions (Sihag *et al.*, 2005 and Abbasi *et al.*, 2007). Its release from organic matter depends upon its source and state of decomposability (Khalil *et al.*, 2001). Organic matter must be broken down by soil microbes to release nitrogen in forms that plants can use. In general, conditions that are good for microbial activity such as warm and moist soil, will promote more rapid breakdown of organic nitrogen. However, some organic nitrogen is in the forms that are difficult for microbes to digest and release under any circumstance. For this reason, it is hard to predict how much and when plant available nitrogen will be released from organic matter.

Many biotic and abiotic factors such as moisture, temperature and nutrient addition drastically affect the decomposition of organic materials added to soil. Soil climatic conditions vary in different ecosystems and seasons, which in turn affect the soil microbial activity and biomass (Brady and Weil, 1999). Soil organic matter decomposition and nitrogen transformation depend on the quality and quantity of organic matter, soil conditions, nutrient status, temperature, moisture, *etc.* (Singh and Kashyap, 2007). The major factors that limit nitrogen transformation in soil are climatic condition such as temperature and soil moisture (Owen *et al.*, 2003 and Lodhi *et al.*, 2009). However, nitrogen transformation processes from a specific organic manure are controlled largely by soil water status, soil type and soil temperature (Sistani *et al.*, 2008; Lodhi *et al.*, 2009). Among these, soil moisture content has great impact on the decomposition and transformation of nutrients (Doel *et al.*, 1990) and is the most important factor that regulates nitrogen transformation (Khalil *et al.*, 2001; Wang *et al.*, 2006; Guntinas *et al.*, 2012 and Chen *et al.*, 2012)

Understanding the forms of nitrogen and how they change within the soil will help to maximise nitrogen supply potential of manure for crop production. Generally, the knowledge of patterns of nitrogen transformation in relation with soil moisture is necessary to synchronize crop nitrogen demands with plant-available nitrogen in the soil.

Keeping the above information in view, the present research programme entitled "*Nitrogen Transformation in Soil Amended with Organic Manures at Different Moistures Regimes*" was conducted with the objective - to study the changes in different forms of inorganic and organic nitrogen in soil amended with organic manures at different moisture regimes.

Materials and Methods

Description of Study Site

An acid soil having the general characteristic as presented in Table 2.1 was collected from Gangapat field of Kongba, Imphal East district, by taking several thin slices from the surface soil layer (0-15 cm) as outlined by Jackson (1973). The composite soils were air dried in shade, ground and passed through 2mm sieve. The processed soils were stored in polyethylene bags and used for the experiment.

Organic manures from different sources *viz.,* farmyard manure (FYM), vermicompost (both from paddy straw and vegetable waste) and biogas slurry were collected and air dried, ground and passed through 2mm sieve. The total nitrogen content of the organic manures are FYM–1.1 per cent, vermicompost (paddy straw)–1.3 per cent, vermicompost (vegetable waste)–1.4 per cent and biogas slurry–1.5 per cent, respectively.

Table 2.1: General Characteristics of the Soil Used in the Experiment

Soil Characteristics	*Results*
Textural class	Silty Clay
Sand (per cent)	11.20
Silt (per cent)	40.00
Clay (per cent)	48.80
Maximum water holding capacity (per cent)	59.63
pH (1:2.5 soil:water ratio)	5.20
EC (1:2.5 soil:water ratio, dsm^{-1})	0.21
CEC [cmol(p+) kg^{-1}]	13.40
Organic carbon (per cent)	1.20
Total nitrogen (per cent)	0.21
Available nitrogen (mg kg^{-1})	154.02
Available phosphorus (mg kg^{-1})	12.50
Available potassium (mg kg^{-1})	64.73
Exchangeable NH_4^+ (mg kg^{-1})	91.50
Soluble NO_3^- (mg kg^{-1})	35.00
Hydrolysable NH_4^+ (mg kg^{-1})	255.00
Hexosamine N (mg kg^{-1})	87.00
Amino acid N (mg kg^{-1})	375.50
(Serine+threonine) -N (mg kg^{-1})	204.00
Unidentified hydrolysable organic N (mg kg^{-1})	557.59
Total hydrolysable organic N (mg kg^{-1})	1275.00
Non-hydrolysable organic N (mg kg^{-1})	177.65

Experimental Treatments and Design

a) Treatment Details

Organic N Sources

T1 = Soil (Control)

T2 = Soil + Farmyard manure (FYM) @ equivalent to 50 mg N kg^{-1}soil

T3 = Soil + Vermicompost (paddy straw) @ equivalent to 50 mg N kg^{-1}soil

T4 = Soil + Vermicompost (vegetable waste) @ equivalent to 50 mg N kg^{-1}soil

T5 = Soil + Biogas manure @ equivalent to 50 mg N kg^{-1}soil

Soil Moisture Regimes

W1 = 40 per cent WHC (Water holding capacity)

W2 = 60 per cent WHC (Water holding capacity)

W3 = 80 per cent WHC (Water holding capacity)

Soil samples were periodically collected and analyzed on 0th, 15th, 30th, 45th, 60th, 75th and 90th days of incubation to determine exchangeable NH_4^+, soluble NO_3^-, total N, total hydrolysable organic N, hydrolysable organic NH_4^+, hexosamine N, amino acid N and (serine + threonine) – N.

b) Layout and Design

The experiment was carried out under factorial randomized block design (FRBD). Altogether there were 15 treatment combinations replicated thrice.

Soil Sampling and Data Analysis

Ten g of air dried soil was taken in each of a series of plastic beakers. Farmyard manure (FYM), vermicompost (paddy straw), vermicompost (vegetable waste) and biogas manure at the rate equivalent to 50 mg N kg^{-1}were mixed thoroughly with the soil according to different set of treatments. Soils of each treatment were moistened to 40 per cent, 60 per cent and 80 per cent of water holding capacity (WHC) and kept at room temperature throughout the experiment. The loss of moisture were replenished by the periodic addition of sterile distilled water on every alternate day by difference in weight. The beakers were kept covered with black polythene sheet and incubated for a period of 90 days. Separate sets of treatments were maintained for each of the sampling stages. Following methods were adopted for analysis of different parameters in the soil samples:

Soil Property	*Methods of Analysis*
Mechanical Analysis	Jackson, 1973
Maximum Water Holding Capacity	Saha, 2008
Soil pH	Jackson, 1973
Electrical Conductivity	Jackson, 1973
Cation Exchange Capacity	Jackson, 1973
Organic Carbon	Jackson, 1973
Available Nitrogen	Jackson, 1973
Available Phosphorus	Jackson, 1973
Available Potassium	Jackson, 1973

Soil Property	*Methods of Analysis*
Total Nitrogen	Jackson, 1973
Exchangeable NH_4^+ and NO_3^- - N:	Hesse, 1971
Organic Nitrogen Fractions	Bremmer, 1965

Statistical Analysis

Data of all the inorganic and organic forms of nitrogen at each stages were statistically analysed through analysis of variance technique for comparing the treatments effects as described by Gomez and Gomez (1984). The significance of various effects was tested at 5 per cent level of probability.

Brief Findings

Nitrogen is one of the most important plant nutrient required by plants for growth and metabolic activities. N undergoes transformation in soil and during the transformation, it is converted from one form to another. The process is influenced by organic matter addition as well as soil moisture. It is difficult to predict the pattern and amount of nitrogen release from organic nitrogen sources during the growing season. Nitrogen availability from applied nitrogen sources must be known for efficient management of nitrogen inputs. Considering the above points, a laboratory incubation experiment entitled "Nitrogen Transformation in Soil Amended with Organic Manures at Different Moisture Regimes" of 90 days was carried out to study the changes in different forms of inorganic and organic nitrogen in soil amended with organic manures at different moisture regimes. The salient findings of the investigation are summarized and concluded here under:

Results showed that in general, changes in the accumulation of different inorganic as well as organic forms of nitrogen were affected greatly by soil moisture regime and organic matter addition. Ammonification, as measured by NH_4^+-N production decreased with increase in moisture regimes in the period of incubation. There was greater accumulation of NH_4^+-N at 40 per cent WHC than other moisture levels. However, nitrification, as measured by NO_3^-N production, generally increased in the whole period of incubation. Increasing soil moisture significantly increased NO_3^-N production and greater accumulation of NO_3^-N was observed at 80 per cent WHC. This may be due inhibition of microbial activity by reducing diffusion of soluble substrate and microbial mobility at low soil moisture. Again, results revealed that comparing with initial value, higher concentration of hydrolysable organic N fractions were observed at the end of the experiment under different moisture levels. Further it was reported that accumulation of hydrolysable organic N fractions including hydrolysable organic NH_4^+, amino acid N, hexosamine N and (serine+threonine)-N were significantly higher at 80 per cent WHC over 40 per cent and 60 per cent WHC revealing that higher moisture favours the production of these organic N fractions. There was accelerated mineralization at 40 per cent WHC showing comparatively lesser amount of organic N during the investigation. This showed that addition of water influenced the transformation of other forms of N to these organic N fractions throughout the incubation period to a greater extent. However, total hydrolysable organic N decreased gradually with increase in incubation time but the accumulation of total hydrolysable nitrogen was significantly higher at medium moisture level *i.e.*, 60 per cent WHC than 40 per cent and 80 per cent WHC. The relatively decrease in total hydrolysable nitrogen supports the observation that hydrolysable N is more vulnerable to mineralization and could be considered as a major sources of potentially available N for plants than non-hydrolysable N. Among the hydrolysable N fractions, unidentified hydrolysable organic N gradually decreased with increase in the period of experiment indicating that this fraction mainly contributes to N mineralization. Whereas significantly higher amount of total nitrogen and other organic nitrogen fractions (unidentified hydrolysable organic N and non-hydrolysable organic N) were found at 40 per cent WHC. Variable trends of different inorganic and organic fractions of N showed that these N forms were affected by moisture regimes.

Further study revealed that, irrespective of different moisture regimes, significantly higher amount of all the inorganic and organic N fractions were accumulated in organic manure amended soils over untreated systems. Among the different organic manures, comparatively higher amount of all the different fractions of N were accumulated in vermicompost (paddy straw) treated soil. This could be attributed higher release of N than other organic manures with smaller losses of N and building of a higher concentration of total N. In the present study, the process of transformation seems to have a conversion of one form of N to other form. The occurrence of different concentration of inorganic and organic forms of nitrogen at different organic manures treated soil showed that the chemical composition, stability and nitrogen content of organic manures favours transformation of nitrogen in soil.

Conclusion

Based on the results obtained from the investigation carried out, it could be concluded that the extent of changes of different inorganic and organic N fractions varied with N fractions, organic matter sources and moisture regimes. Application of organic manures is an effective way of maintaining the quantity and quality of soil inorganic and organic N. Among the organic manures, vermicompost (paddy straw) not only add N to soil but also promotes mineralization of N in soil. The chances of N loss from the farming system will be reduced thereby increasing availability to crop. The traditional use of manures should therefore, be promoted in order to maintain long term soil productivity. Changes in soil moisture due to management of water resources will affect N transformation. Lower moisture favours N mineralization in the soil. With the proper addition of organic manures at suitable moisture regimes, it may be possible to maintain a constant pool of available Nitrogen. So, before using different sources of organic manures in the field due consideration should be given on N release pattern of those organic manures as well as moisture status

REFERENCES

1. Sihag, D., Singh, J.P., Mehla, D.S. and Bhardwaj, K.K. (2005). Effect of Integrated Use of Inorganic fertilizers and Organic Materials on the Distribution of Different forms of Nitrogen and Phosphorus in Soil. *J. Indian Soc. Soil Sci.*, **53**(1): 80-84.
2. Abbasi, M.K., Hina, M., Khalique, A. and Khan, S.R. (2007). Mineralization of three organic Manures used as Nitrogen source in a soil Incubated under Laboratory Conditions. *Commun. Soil Sci. Plant Anal.* **38**(13-14): 1691-1711.
3. Khalil, M.I., Cleemput, O.V., Boeckx, P. and Rosenani, A.B. (2001). Nitrogen transformation and emission of greenhouse gases from three acid soils of humid tropics amended with nitrogen sources and moisture regime. 1. Nitrogen transformations. *Commun. Soil Sci. Plant Anal.* **32**(17-18): 2893-2907.
4. Brady, N.C. and Weil, R.R. (1999). The nature and Properties of Soils, Prentice Hall, Upper Saddle River, New Jersey.
5. Singh, J.S. and Kashyap, A.K. (2007). Variation in soil N – mineralization and nitrification in seasonally dry tropical forest and savanna ecosystems in Vindhyan region, India. *Trop. Ecol.* **48**(1): 27-35.
6. Owen, J.S., Wang, M.K., Wang, C.H., King, H.B. and Sun, H.L. (2003). Net nitrogen mineralization and nitrification rates in a forested ecosystem in north eastern Taiwan. *For. Ecol. Manag.* **176**(1-3): 519-530.
7. Lodhi, A., Arshad, M., Azam, F. and Sajjad, M.H. (2009). Changes in Mineral and Mineralizable N of soil incubated at varying salinity, moisture and temperature regimes. *Pak. J. Bot.* **41**(2): 967-980.
8. Sistani, K.R., Adeli, A., McGowen, S.L., Tewolde, H. and Brink, G.E. (2008). Laboratory and field evaluation of broiler litter nitrogen mineralization. *Bioresour. Technol.* **99**: 2603-2611.
9. Doel, D.S., Honeycutt, C.W. and Halteman, W.A. (1990). Soil water effects on the use of heat units to predict crop residue carbon and nitrogen mineralization. *Biol. Fertil. Soils.* **10**: 102-106.
10. Wang, C., Wan, S., Xing, X., Zhang, L. and Han, X. (2006). Temperature and soil moisture interactively affected soil net N mineralization in temperate grassland in Northern China. *Soil Biol. Biochem.* **38**: 1101-1110.
11. Guntinas, M.E., Leiros, M.C., Trasar-Cepada, C. and Gil-Sotres, F. (2012). Effects of moisture and temperature on net soil nitrogen mineralization: A laboratory study. *Eur. J. Soil Biol.* **48**: 73-80.
12. Chen, Q.H., Feng, Y., Zhang, Y.P., Zhang, Q.C., Shambi, I.H., Ying, F., Yan-Ping, Z., Qi-Chun, Z., Zhang, Y. and Lin, X.Y. (2012). Short - Term Response of Nitrogen

Mineralization and Microbial Community to Moisture Regimes in Greenhouse Vegetable Soils. *Pedosphere.* **22**(2): 263-272.

13. Saha, A.K. (2008). Methods of Physical and Chemical Analysis of Soil. Kalyani Publishers, New Delhi.
14. Bremner, J.M. (1965). Organic forms of nitrogen. In: Black, C.A. *et al.* (eds) Methods of Soil Analysis. *Am. Soc. Agron.*, USA 9, Part **2**, 1238-1253.
15. Jackson, M.L. (1973). Soil chemical analysis. Prentice Hall of India Pvt. Ltd., New Delhi.
16. Hesse, P.R. (1971). A textbook of soil chemical analysis. John Murray (Publishers) Ltd., London.
17. Gomez, K.A. and Gomez, A.A. (1984). Statistical procedures for agricultural research. John Wiley and sons. Inc. London, UK (2nd edtn) pp. 13-175.

Section B

Nutrient Management in Rapeseed

Chapter 3

Integrated Nutrient Management in Rice-Mustard Cropping Sequence

W. Herojit Meetei

ABSTRACT

A field experiment was conducted for two consecutive years to study the effect of integrated nutrient management (INM) on nutrient release pattern, uptake and yield in rice (var. CAUR-1)-mustard (toria var. M-27) cropping sequence. An incubation study was also done to study the effect of INM on nutrient release pattern under different moisture regimes (40, 60 and 80 per cent WHC). Urea, Vermicompost, Azolla and Azotobacter were applied as nitrogen sources including one untreated control comprising twelve treatment combinations. Application of the various organic and inorganic sources of nutrients affected nutrient release pattern, their uptake, plant growth and yield attributes significantly. Result revealed that INM markedly improved the available nutrient content, their uptake, plant height, number of tillers, effective tillers hill-1, panicle length, test weight, grain and straw yield for rice over single inorganic application and untreated control. Significantly higher available N, P, K, Zn and Mn content in rice soil were found in T4 which is at par with T11 and T6. Comparing among the different treatments, the highest N, P, Zn, Fe, Mn and Cu uptake was observed in soil added with T4 which was at par with T6. Soil treated with T4 yielded significantly higher grain yield which was followed by T11 showing parity with T6. Higher nutrient uptake resulted in higher production. Integrated nutrient management also significantly enhanced plant height, siliquae plant^{-1}, seeds siliqua^{-1} and test weight, stover and seed yield of toria var. M-27 over the sole application of inorganic source. The effect of application of T11, T8 and T6 were similar on macronutrients (N, P and K) and micronutrients (Fe, Mn and Cu) availability in soil grown with mustard as well as incubation study. Irrespective of different treatments, there was an increasing trend of macro (N, P and K) and micronutrients (Zn, Fe, Mn and Cu) uptake up to

90th day after sowing (DAS) of mustard crop followed by a decline at harvest. Comparatively higher macro and micronutrients uptake by mustard as well as seed yield were recorded in T8 added soil at harvest showing parity with T11. Positive and significant correlation of yield and N uptake of rice and mustard with different organic N fractions showed the contribution of these fractions to available N pool. Result of incubation study revealed that in general higher moisture favours nutrient availability in the soil. Integrated nutrient management improves soil fertility while sustaining crop productivity.

Note: T4 -(75 per cent RDN using Urea + 25 per cent RDN using Vermicompost), T6 - (50 per cent RDN using Urea + 50 per cent RDN using Vermicompost), T8- (25 per cent RDN using Urea + 75 per cent RDN using Vermicompost) and T11- (100 per cent RDN using chemical fertilizer Urea + 100 per cent RDN using Vermicompost + 100 per cent using *Azolla*).

Introduction

Agriculture is and will be the backbone of Indian economy contributing to a significant part of the gross domestic product where two thirds of the population is engaged either directly or indirectly in farm related activities. Agriculture is an art and science of growing crops that is satisfying the most fundamental need of man *i.e.*, food has influenced the social and economic life of the people of India for centuries.

At the beginning of new millennium, the evidence is overwhelming that an agricultural transformation is essential to meet the global challenges of feeding the world's growing population, conserving the environment and reducing poverty. Such a transformation is urgently needed in large agrarian economies such as India where it will have to occur at the level of small-holder farmers so that their complex farming systems can be made more productive and more efficient in the use of resources (Serageldin, 2000).

India achieved self-sufficiency with rapid growth and development in food grain production from a mere 51.0 Mt in fifties to 275.68 Mt now (Kumari, 2020). This was possible through co-operative effort made by farmers and scientists by bringing the research from laboratory to land. There has been number of reports from almost every state in the country that in spite of increased input the crop production has either levelled off or declined because of reduction in organic carbon content of soil. Soils are being constantly degraded and destroyed with profound economic costs. About 70 per cent of soils are deficient in organic carbon (less than 1 per cent) and micronutrients deficiencies are being widely experienced throughout the country. Soil toxicity due to use of chemicals, pesticides and industrial effluents is affecting adversely both soil health and crop productivity (Paroda, 1998). The results generated from series of long-term fertilizer experiments revealed that continuous use of high analysis chemical fertilizers increased the crop yield in the initial years but adversely affected the sustainability yet a later stage (Virmani, 1994). Indiscriminate use of high analysis chemical fertilizers also resulted in the deficiency of nutrients other than the applied and caused decline in organic carbon (Singh *et al.*, 1999). However, when manure was applied in conjunction with chemical fertilizers, decline in organic carbon was arrested and the gap between potential yield and the actual yield is bridged largely. Therefore, use of chemical fertilizers alone may not keep pace with time in the maintenance of soil health and sustaining the productivity.

Rice (*Oryza sativa* L.) is a staple food crop for a large proportion of world's population, and its demand is accelerating with increasing population (Bezbaruha *et al.*, 2011). Rice has a unique position in Indian economy. It accounts for almost 22 per cent of the total world paddy production (Varma, 2017). India is the world's second largest rice producer having an area of 43.7 million hectares contributing approximately 45 per cent of total cereal production and 42 per cent of total food grains (Mathur and Krishraiah, 1996). However, the production is very low 88.6 million tons as compared to other rice growing countries of the world. Per capita consumption averages to around 86 kg per year. In Manipur, rice is grown in both the hill and plain areas, cultivated on 168.78 thousand hectare, with an annual production of 3.87 lakh tonnes. Among the districts, Thoubal District had the highest production of rice with 70.97 lakh thousand and lowest was recorded in Chandel District having only 18.69 thousand tonnes during the year 2011-12. The major constraints in obtaining higher crop yields in this region are vagaries of nature, low fertilizers use efficiency, partial implementation of recommended management practices and heavy infestation of weeds. In the meantime, prevailing high soil acidity has also practically discouraged the use of phosphate fertilizers, which quickly get fixed into insoluble forms.

India holds a premier position in global oil seeds scenario accounting for 19 per cent of the total oil seed area and 9 per cent of oil seed production; however, the productivity in India is only 1086 kg ha^{-1}as compared to world level of 1632 kg ha^{-1} (Yaspal *et al.*, 2014). Edible oils have an important role to play in relieving malnutrition and calories requirements of human and animal population. Now, the total oilseed production as per 2nd Advance Estimates during 2015-16 is projected to be around 25.25 million tonnes (https://pib.gov.in/PressReleasePage.aspx?PRID=1482805, accessed on 23-03-2022) and the country may have to import more vegetables oils in the next decade due to increase in population and improved in the standard of living. India would need 58 million tonnes of oilseeds by 2020 for maintaining minimum edible oil requirement of 12 kg/capita/annum (Mittal, 2008) from the present level of 8.2 kg. With the present day levels of productivity (980 kg ha^{-1}), we are very far from the target to be achieved.

India is the fourth largest producer of oilseeds accounting for about 19 per cent of the global area, 2.7 per cent of global production in the world (Reddy and Immanuelraj, 2017). Among various oilseed crops grown, rapeseed-mustard occupies prominent position next to the groundnut and soybean. At present, production of oilseeds in India is around 29.80 million tonnes from 26.31 million hectare with an average productivity of 1133 kg ha^{-1}. (Jat *et al.*, 2019).

However, yields are still low due to poor adoption of agronomic practices, of which nutrient management is most important component. Mustard is an exhaustive and energy rich crop, hence responds well to nutrient application. Field experiments conducted at various locations in the country showed linear response of mustard to nitrogen fertilization up to 80 kg ha^{-1}(Singh *et al.*, 1998) and 100 kg ha^{-1}(Aulakh and Pasricha, 1998) and 120 kg ha^{-1}(Patil *et al.*, 1996). Depending on soil moisture and available phosphorus, response of mustard to phosphorus varied from 30 to 60 kg P_2O_5 ha^{-1} (Narang *et al., 1993*). In Manipur, the total production of mustard during 2016-17 was 249060 quintals as against the 245200 quintals in 2015-16 showing an increase of 3860 quintals over that of the previous year (Devi *et al.*, 2017).

Mustard being a tap-rooted crop with well spread lateral root system can be grown both under rainfed and irrigated conditions. It offers good opportunity for diversification of cropping pattern in *rabi* season in Manipur where rain is scanty during winter. The production levels are below the regional and national averages, providing ample scope for increasing both production and productivity. The principal causes of low productivity for obtaining higher crop yields in this region are prevailing high soil acidity, late sowing time, low to very low utilization of nitrogenous and phosphate fertilizers. The importance of fertilizer is increasing the production of food grains and other crop is well recognized by the farmers. However, under intensive cropping neither the chemical fertilizers nor the organic source of nutrients alone can sustain higher productivity. It has also been observed in long term studies that higher yield level could not be maintained even by the use of balanced NPK fertilizers. This has been attributed to deterioration of soil physical conditions and emergence of other macro and micronutrients deficiencies. The unsustainability of higher yield is further accentuated by increasing cost of fertilizers.

The continuous application of fertilizers alone in large quantities over a long period in intensive cropping system results in imbalance in supply of other nutrients and adverse effect on physical properties and chemical composition of soil, which finally leads in declining crop yields. The integrated use of organic manures, inorganic fertilizers and bio-fertilizers can help to maintain optimum crop yields stability through correction of marginal deficiencies of secondary and micronutrients, enhancing of applied nutrients, providing favourable soil physical condition and required soil nutrient pool on sustained basis. Integrated plant nutrient approach is going to be the mainstay in the next millennium. This concept is closely related to that of FAO's Integrated Plant Nutrient System (IPNS), where the basic goal is the maintenance or adjustment of soil fertility and of plant nutrient supply to an optimum level for sustaining the desired crop productivity by optimization of the benefits from all possible sources of plant nutrients in an integrated manner.

Integrated nutrient management (INM) is a prescription for developing a durable IPNS. Integrated nutrient management (INM) aims at efficient and judicious use of all the sources of plant nutrients in an integrated manner, so as to attain sustainable crop production with minimum deleterious effect of chemical fertilizers on soil health and least disturbance to the plant-soil environment (Pillai, 1996), arrest degradation of soil, water and environment quality by promoting carbon sequestration, minimizing the avoidable leakage of the fertilizer nutrients to water bodies and atmosphere. There is a wide scope for increasing nutrient supply through use of organic manures, bio-fertilizers and adoption of proper crop sequences and these together can contribute significantly to the required nutrient pool. However, there is no scope for reducing the consumption of inorganic fertilizers since the level of crop productivity is not only to be maintained but is to be increased in the coming years. Under the present Indian context, the organic sources of nutrients should be considered as supplements to chemical fertilizers and not their substitutes (Mishra and Kapoor, 1992) and this is true particularly for the state of Manipur.

Normally only 35 to 40 per cent of N from organic manure added is available for the use of crop. Rest of it is available to the succeeding crops. Hence, the beneficial effect of cropping sequence is improvement of soil health and fertility of soil due to addition of organic manure such as vermi-compost. However, the change over from chemical to organic agriculture is difficult, as farmers have to forgo yields till the build-up of soil productivity to a desirable level. But, it is possible to induce a quick change over to sustainable agriculture by harnessing different integrated nutrient management practices mainly bio-fertilizers, vermi-compost, and micro-organism.

Abrol and Gill (1994) have identified three soil factors related to unsustainability of crop production system; namely loss of organic matter, diminishing trend in nutrient supplying capacity of soil and impairment of physical and physico-chemical properties of soil. This underscores the importance of using adequate quantities of organic sources of nutrient in conjunction with inorganic fertilizers in an integrated manner following a management technology that is efficient, economically viable, socially acceptable and ecologically sound. Keeping all these facts in view, the present study on "Integrated nutrient management practices in rice – mustard cropping sequence" was planned during the *kharif* and *rabi* seasons of 2015 and 2016 at the Research Farm, College of Agriculture, Central Agricultural University, with the following objectives:

1. To study the effect of integrated nutrient management on soil properties
2. To study the release patterns of nutrients and
3. To study the effect of integrated nutrient management on nutrient uptake and yield of rice and mustard.

Materials and Methods

Description of Study Site

The experiments conducted during both the years were laid at the Research

Farm of the Department of Soil Science and Agricultural Chemistry, College of Agriculture, Central Agricultural University, Imphal, Manipur. The site was situated at 24°48'44.50"N latitude, 93°53'29.98"E longitude and at an altitude of 790m above mean sea level. In general, the Imphal valley falls under sub-tropical climate. The average annual rainfall of this tract is about 1467.50 mm. Monsoon rain starts from the month of June and continues up to October with pre-monsoon showers from March onwards.

Experimental Treatments and Design

The present investigation was conducted at the experimental field for two consecutive years during *Kharif* and *Rabi* season of 2015-2016 and 2016-2017. In *Kharif* season, rice (*Oryza sativa* L.) and in *Rabi* season, mustard (*Brassica sp.*) was grown.

Experimental Design	*Randomized Block Design*	
Number of replications	3	
Number of treatments	12	
Plot size	4 m X 5 m	
Border between the replications	50 cm	
Border between plots	30 cm	
Gross experimental area	900 m^2	
Net experimental area	720 m^2	
Spacing	*Row-Row (cm)*	*Plant-Plant(cm)*
a) Rice	20	10
b) Mustard	20	5
Total number of plots	36	
Recommended dose of fertilizer a) Rice	60:40:30 NPK Kg ha^{-1}	
b) Mustard	40:30:20 NPK Kg ha^{-1}	

Treatment Details

There were 12 (twelve) treatments. The recommended doses of nitrogen (RDN) *i.e.* 60 kg N ha^{-1} for paddy and 40 kg N ha^{-1} for mustard were met from urea, vermicompost and *Azolla* applied either singly or in combination as per treatment except control which received no nitrogen. Vermicompost and *Azolla* contained 1.4 per cent N and 3.44 per cent N (dry weight basis), respectively. A constant dose of 40 Kg P_2O_5 ha^{-1} and 30 Kg K_2O ha^{-1} for paddy and 30 Kg P_2O_5 ha^{-1} and 20 Kg K_2O $ha^{-1}$for mustard were applied through SSP and MOP as basal in all the plots.

1. Treatment Details of Rice (*Kharif* season)

Notations	*Treatment Combinations*
T0	Control
T1	100 per cent RDN using chemical fertilizer Urea
T2	100 per cent using Vermicompost as N-source
T3	100 per cent using *Azolla* as N-source
T4	75 per cent RDN using Urea + 25 per cent RDN using Vermicompost
T5	75 per cent RDN using Urea + 25 per cent using *Azolla*
T6	50 per cent RDN using Urea + 50 per cent RDN using Vermicompost
T7	50 per cent RDN using Urea + 50 per cent using *Azolla*
T8	25 per cent RDN using Urea + 75 per cent RDN using Vermicompost
T9	25 per cent RDN using Urea + 75 per cent using *Azolla*
T10	00 per cent RDN using Vermicompost + 100 per cent using *Azolla*
T11	100 per cent RDN using chemical fertilizer Urea + 100 per cent RDN using Vermicompost + 100 per cent using *Azolla*

2. Treatment Details of Mustard (*Rabi* season)

Notations	*Treatment Combinations*
T0	Control
T1	100 per cent RDN using chemical fertilizer Urea
T2	100 per cent using Vermicompost as N-source
T3	100 per cent using *Azolla* as N-source + *Azotobacter*
T4	75 per cent RDN using Urea + 25 per cent RDN using Vermicompost
T5	75 per cent RDN using Urea + 25 per cent using *Azolla* + *Azotobacter*
T6	50 per cent RDN using Urea + 50 per cent RDN using Vermicompost
T7	50 per cent RDN using Urea + 50 per cent using *Azolla* + *Azotobacter*
T8	25 per cent RDN using Urea + 75 per cent RDN using Vermicompost
T9	25 per cent RDN using Urea + 75 per cent using *Azolla* + *Azotobacter*
T10	100 per cent RDN using Vermicompost +100 per cent *Azolla* + *Azotobacter*
T11	100 per cent RDN using chemical fertilizer Urea + 100 per cent RDN using Vermicompost + 100 per cent using *Azolla* + *Azotobacter*

Soil Sampling and Data Analysis

The soil samples were randomly collected from different spots of experimental site up to 0-20 cm depth before the start of the experiment and a homogenous composite sample was prepared after proper mixing, drying and sieving. The composite soil sample were analyzed for different physico-chemical characteristics and nutrient status of the experimental field. The results of the analysis are presented in Table 3. Five soil samples were collected randomly from each plot from the surface soil layer (0- 20 cm) and mixed to make one composite sample. The composite samples were immediately transported to the laboratory and stored at 4°C for analysis of inorganic N and one sub sample was air dried at room temperature for analysis of the remaining parameters (macro and micro nutrients and nitrogen fractions). Sampling was done periodically at regular intervals of 0th, 30th, 60th and 90th DAT and at harvest for rice and 0th, 30th, 60th and 90th DAS and at harvest for mustard.

Table 3. Physico-Chemical Characteristics of Soil (0-20 cm)

Soil Property	*Result*	*Interpretation*	*Methods of Analysis*
Textural class		Clayey soil	Bouyoucos Hydrometer method (Jackson, 1973)
Sand (per cent)	8.24		
Silt (per cent)	13.44		
Clay (per cent)	78.32		
Bulk density (g cm^{-3})	1.43	-	Soil core method (Jackson, 1973)
pH (1:2.5)	5.4	Moderately acidic	1:2.5 soil: water suspension (Jackson, 1973)
EC (dSm^{-1})	0.28	Normal	1:2.5 soil: water suspension (Jackson, 1973)
CEC [Cmol(p+)kg^{-1}]	14.05	Medium	Leaching with 1N NH_4OAc (Jackson, 1973)
Organic carbon (per cent)	1.56	High	Walkley and Black (1934)
Ca [Cmol(p+)kg^{-1}]	5.05	Low	EDTA Titration method (Jackson, 1973)
Mg [Cmol(p+)kg^{-1}] 38	4.96	Low	
Available Nitrogen (Kg N ha^{-1})	9.39	Medium	Alkaline potassium permanganate method (Subbiah and Asija, 1956)
Available Phosphorus (Kg P_2O_5 ha^{-1})	57.71	High	Bray and Kurtz No.1 Method (Jackson, 1973)
Available potassium (Kg K_2O ha^{-1})	252.67	Medium	Flame photometer (Jackson, 1973)
Total Nitrogen (per cent)	0.12	Poor	Modified macro-Kjeldahl (Jackson, 1973)
DTPA extractable Zn (mg kg^{-1})	0.80	Sufficient	DTPA method (Lindsay and Norvell, 1978)

Soil Property	*Result*	*Interpretation*	*Methods of Analysis*
DTPA extractable Fe (mg kg^{-1})	94.38	Sufficient	
DTPA extractable Mn (mg kg^{-1})	35.81	Sufficient	
DTPA extractable Cu (mg kg^{-1})	0.96	Sufficient	

Plant Sample Analysis

Plant samples were collected from each plot and oven dried at 65 0C for 72 hours. The collected samples were ground and used for determination macro (N, P and K) and micro (Zn, Fe, Mn and Cu) nutrients content following the standard procedures as mentioned below.

Methods Used for Determination of Nutrients in Plant Sample

Element	*Method Employed*	*Reference*
N	Modified Micro-Kjeldahl	Jackson, 1973
P	Wet digestion method and P was estimated by Vanadomolydophosphoric yellow colour	Jackson, 1973
K	The extract from the wet digest was used for determining Potassium using flame photometry	Jackson, 1973
Zn	Atomic absorption	Elwell and Gridley, 1967
Fe	Atomic absorption	Elwell and Gridley, 1967
Mn	Atomic absorption	Elwell and Gridley, 1967
Cu	Atomic absorption	Elwell and Gridley, 1967

Statistical Analysis

Data obtained from the experiments were statistically analysed through analysis of variance technique for comparing the treatments effect as described by Gomez and Gomez (1984). The significance of various effects was tested at 5 per cent level of probability. Correlation studies were calculated using SPSS software version 16.0.

Brief Findings

Application of inorganic and organic sources of nitrogen either singly or in combination significantly affected nutrient availability, changes in nitrogen fractions, nutrient uptake, plant height, number of effective tillers $hill^{-1}$, panicle length, test weight and yield (grain and straw) of paddy. The result revealed that irrespective of different treatments, available nitrogen content in paddy soil increased up to 30th days after transplanting (DAT) and then gradually decreased in each successive sampling days *i.e.* 60th and 90th DAT and at harvest in both the consecutive years and pooled data. However, available phosphorus and potassium increased up to 60th DAT followed by a decline at harvest. Irrespective of different treatments DTPA extractable zinc and copper content in nitrogen applied paddy soil gradually declined till harvest. Available iron and manganese content in soil increased up to 30th DAT followed by a gradual decline till harvest. The declining trend of available nutrients with crop growth might be due to plant uptake. An initial increase in available nutrient might be attributed to the fact that application of organic sources enhanced the rate of mineralization resulting rapid release of nutrient. All the treatments involving INM significantly influenced the available macro and micronutrient content as compared to control at different stages of crop growth. The increase in available nutrients in soil with the application of INM could be due to better microenvironment with efficient mobilization of mineral nutrients and effect of applied organic sources. Further perusal of the data revealed that comparing among the different treatments, significantly higher available nitrogen, phosphorus, potassium, zinc and manganese concentration were recorded in T4 which is at par with T11 and T6. There was comparatively higher accumulation of available copper in soil treated with T11 which was statistically at par with T8. Also, pooled mean data revealed that T4 accumulated significantly higher available Fe followed by T11 which showed similarity with T6 at the time of harvest.

Irrespective of different treatments, the amount of exchangeable NH_4^+, soluble NO_3^-, hydrolysable organic NH_4^+, total hydrolysable N and total N content in soil gradually decreased up to harvest. The decrease in inorganic forms of N might perhaps be due to conversion to other forms, clay fixation, immobilization, volatilization and leaching or plant uptake. The decline in total hydrolysable N with time supports the observation that this fraction is more vulnerable to mineralization and could be considered as a major source of potentially available N for plants. Irrespective of different treatments and year of study the amount of amino acid decreased up to 30th DAT followed by a slight increase up to 90th DAT with a decline at harvest. The decrease in hydrolysable NH_4^+ and amino acid nitrogen in soil might be because of the mineralization of these fractions of organic N under favourable microclimate thereby contributing available nitrogen. Similar trend of increment in (serine + threonine) and hexosamine N content during the whole period of crop growth was recorded in both second year and pooled data. The increase in the amount of (serine + threonine) and hexosamine N over the whole period of experiment might be due to the conversion of other forms of nitrogen to these organic forms. In general, addition of inorganic or organic N sources and their combinations significantly increased different nitrogen fractions over untreated control at different stages of crop growth. Furthermore, data revealed that significantly higher concentration of different forms of nitrogen was observed in soil treated with integration of various N sources as compared to the single application of inorganic source. Addition of mineral N along with organic sources narrowed the C: N ratio of organic manure and this enhanced the rate of mineralization resulting rapid release of N from the organic carbon. Further study of the pooled data shown that comparatively higher accumulation of exchangeable

NH_4^+ and soluble NO_3^--N was observed in soil amended with T11 which is at par with T4. Comparing among the different treatments, maximum values of organic nitrogen fractions and total nitrogen content were recorded in T11 which showed parity with T10 and T8.

Regarding plant nutrient concentration, an increasing trend of nitrogen content was noticed up to 60th day after transplanting followed by a decreasing trend till harvest stage. Result further revealed a decreasing pattern of phosphorus, potassium and micronutrients (zinc, iron, manganese and copper) content in rice with increase in crop age in both the years which might be due to dilution effect caused by higher dry matter production, low nutrient status of soil and fixation of applied nutrient. Application of integrated nitrogen management improved the N, P, K, Zn, Fe, Mn and Cu content of rice over untreated control in both the years of study as well as pooled data. Statistically higher nutrient concentration in rice was observed in soil treated with T4 which was statistically at par with T6. Irrespective of different treatments, dry matter yield of paddy increased with advancement of crop growth till harvest stage. The pattern of nitrogen, phosphorus, potassium, iron, copper and zinc uptake by paddy showed an increase up to 90th DAT followed by a decline at harvest. However, manganese uptake was different with gradual increase till harvest. Further, it was also observed that the uptake of nutrients by paddy from soils amended with inorganic and organic nitrogen sources had significant effect over control at different stages of crop growth. Comparing among the different treatments, higher nitrogen, phosphorus, zinc, iron, manganese and copper uptake was observed in soil added with T4 which was at par with T6.

Critical study showed that INM treated soil produced significantly more number of effective tillers hill-1, plant height, panicle length and test weight over untreated control. Analysis of the data revealed that significantly higher grain and straw yield of paddy was recorded in soil treated with inorganic and organic sources of nitrogen and their combinations over untreated control. Critical analysis of data of both consecutive years and also pooled mean values showed that T4 yielded significantly higher grain yield which was followed by T11 showing parity with T6 and T2. Similarly, maximum straw yield was recorded in T4 which was at par statistically with T11 and T6. The result showed the efficiency of organic sources of nutrients in production and improving the soil fertility levels.

Result of the effect of INM on soil properties after harvest of paddy revealed that combined application of inorganic and organic sources had significantly higher organic carbon, cation exchange capacity, macro and micro-nutrients availability in soil over control and sole application of inorganic source (T1). Correlation studies showed that there was positive relation between all the fractions of nitrogen and nitrogen uptake (grain and straw) and yield (grain and straw) of rice. The highest correlations of N uptake by grain and straw were found with ammonium-N with r values of r = 0.910** and 0.830**, respectively. Grain and straw yield of rice were highly correlated with ammonium N (r = 0.879** and 0.877**, respectively) followed by nitrate-N (r = 0.793** and 0.781**, respectively).

Single or combined application of organic and inorganic sources of nitrogen affected nutrient release pattern, changes in nitrogen fractions, nutrient uptake, plant growth and yield attributes of mustard significantly. Available nitrogen, potassium and micro-nutrients (zinc, iron, manganese and copper) content in mustard grown soil increased up to 30th days after sowing (DAS) followed by a decline till harvest. However, release of phosphorus increased up to 60th DAS with a slight decline at harvest irrespective of different treatments. The declining trend of available nutrients with crop growth might be due to plant uptake or immobilization by organic sources in soil. Data also revealed significantly higher available nutrients in soil applied with different organic nitrogen sources and their combinations with inorganic source over untreated control. Critical study of the result showed that the effect of application of T11, T8 and T6 were similar on macronutrients (nitrogen, phosphorus and potassium) and micronutrients (iron, manganese and copper) availability in soil. However, application of T8 maintained significantly higher amount of DTPA extractable zinc followed by T11 which was at par with T6. Results further revealed that irrespective of different treatments there was a gradual declining trend of NH_4^+-N, total hydrolysable and total N till the time of harvest. The decline in NH_4^+-N with crop growth might perhaps be due to the transformation of exchangeable NH_4^+-N to NO_3^--N, clay fixation, immobilization or loss through volatilization. The decline in total hydrolysable N with time supports the observation that this fraction is more vulnerable to mineralization and could be considered as a major source of potentially available N for plants. The amount of soluble NO_3^- and hydrolysable NH_4^+-N increased up to 30th day after sowing (DAS) followed by a decline till harvest stage of mustard. Mineralization of organic sources of N increases NO_3^- -N. However, immobilization and plant uptake causes its decline. Hexosamine N content declined up to 60th DAS followed by an increase till harvest. Changes in the amount of amino acid content showed a different pattern of decrease up to 30th DAS followed by a slight increase up to 60th DAS with a decline till harvest. Mineralization of organic fractions might cause the decline and increase might be due to inter-conversion between different forms of N. Further study of the results showed that significantly higher accumulation of inorganic and organic N fractions, total hydrolysable and total N was recorded in all treatments involving N addition over untreated control throughout the study. It was also observed that conjunctive application of inorganic and organic sources of N accumulate maximum amount of all the N fractions and the total N over the sole inorganic addition. Comparing among the different treatments, higher amount of exchangeable NH_4^+ was observed in soil treated with T11 but statistically at par with T6 and T4. Whereas NO_3^- -N content was recorded in soil treated with T11 followed by T4 which was at par with T6 on 90th DAS and at harvest. Similar effect on accumulation of hydrolysable NH_4^+-N, hexosamine, (serine + threonine)-N and total hydrolysable N was found in soil treated with T11, T10 and T8 at the time of harvest. However, soil treated with T11, T6 and T8 maintained statistically similar amino acid content in soil at harvest.

There was an increase in nitrogen content in mustard up to 60th DAS followed by a declining trend till the harvest stage. However, a progressive decline in phosphorus, potassium and micro-nutrients content (zinc, iron, manganese and copper) with advancement of crop growth was observed. The progressive decline of nutrient content with the advancement of crop growth might be due to three reasons: dilution effect, caused by higher dry matter production and fixation of native and applied nutrient. Study of the pooled data revealed that significantly higher macro and micronutrients content was recorded in inorganic and organic sources applied singly or in combination over untreated control. Comparing among the different treatments, there was no significant difference in nitrogen content of mustard among the N sources treatment on 60th DAS. However, analytical higher N content was found in T4 which was statistically at par with T6, T7, T8, T9 and T11 at harvest time. It was also observed that P, K and micronutrients (Zn, Fe and Cu) content in mustard was significantly higher in T8 which exposed similarity with T11 and T4, respectively. Statistically higher Mn accumulation was recorded in T8 showing parity with T11 followed by T4. Increase in dry matter yield of mustard was also recorded with increase in crop growth till harvest stage. Higher dry matter production might be due to increased availability of major nutrients through chemical fertilizer and mineralized nutrient from organic sources throughout the cropping period. Further, data showed that significantly higher dry matter yield of mustard was observed in T8 which was statistically at par with T11, T4 and T6 at different stages of crop growth followed by T1 on 90th DAS and at harvest. Irrespective of different treatments, there was an increasing trend of macro (N, P and K) and micro-nutrients (Zn, Fe, Mn and Cu) uptake up to 90th DAS of mustard crop followed by a decline at harvest. Higher dry matter yield leads to enhancement of proficient growth and development of mustard through highest uptake and utilization of the nutrients. In general, the application of inorganic and organic nitrogen sources either singly or in combination to soil had significant effect on nutrients uptake by mustard over control at different stages of crop growth. Perusal of the data showed that comparatively higher macro (N, P and K) and micronutrients (Zn, Fe, Mn and Cu) uptake by the crop was recorded in T8 added soil at harvest stage but the treatment was at par with T11 in P, K and micronutrients uptake.

Scrutiny of the results pointed out that combined application of inorganic and organic sources of nitrogen significantly enhanced plant height, produced more number of primary and secondary branches plant-1, siliquae plant-1, seeds siliqua^{-1}and test weight over untreated control. The integrated treatments also recorded statistically higher seed and stover yield over sole addition of inorganic or organic N sources and untreated control. Comparing among the integrated treatments, significantly higher seed yield was found in T8 which was statistically at par with T11, whereas in case of stover yield, maximum yield was recorded in T8 followed by T11.

Evaluation of the data on effect of INM on soil properties after harvest of mustard revealed that soil with INM treatments had significantly higher organic carbon, cation exchange capacity, macro and micronutrients availability than inorganic fertilizer treatment and untreated control. Correlation analysis found that seed and stover yield of mustard had given the highest correlation coefficient (r) values with nitrate-N (r = 0.925** and 0.846**, respectively) followed by ammonium-N (r = 0.897** and 0.836**, respectively). The maximum correlations of N uptake by seed and stover were found with nitrate-N (r = 0.928** and 0.883**, respectively). The positive and significant correlation of yield and N uptake of mustard with different organic N fractions showed the contribution of these fractions to available N pool.

Analysis of the data of incubation study revealed that irrespective of different treatments accumulation of available nitrogen, phosphorus, iron and manganese gradually increased up to 60th day of incubation, thereafter continued to decrease slowly up to the last stage of incubation under different moisture regimes. However, there was an increasing trend of available potassium up to 30th day and then, declined till 90th day of incubation. In general, irrespective of different treatments and moisture regimes, DTPA-extractable zinc and copper gradually decreased throughout the incubation period. Irrespective of different inorganic and organic matter treated soils, statistically higher amount of available nitrogen, phosphorus and potassium was found in soil maintained at 60 per cent WHC followed by 80 per cent and 40 per cent WHC. This might be due to higher microbial activity resulting greater mineralization at higher water content than at lower water content. Comparatively greater accumulation of DTPA-extractable zinc and copper was observed at 60 per cent WHC followed by 40 per cent and 80 per cent WHC at different stages of incubation. It signified that DTPA-extractable zinc and copper content in soil decreased with increasing moisture level. Closer examination of the data revealed that significantly greater accumulation of DTPA-extractable iron and manganese was recorded at 80 per cent WHC followed by 60 per cent and 40 per cent WHC at different stages of incubation. Irrespective of different moisture regimes, significantly higher accumulation of macro and micro nutrients was noticed in all the nitrogen treated soils than untreated control at different stages of incubation. Further perusal of the data showed that integration of inorganic and organic N sources resulted significantly higher amount of macro and micro nutrients than inorganic treated soil (T1). Enhanced nutrient content in integrated treatments might be due to gradual release of nutrients from organic sources which increased the nutrients to the soil along with inorganic source and made it available to the soil. Comparing among the treatments, T11 showing parity with T4 and T6 maintained significantly higher accumulation of available nitrogen, potassium and iron in soil throughout the incubation period. Among the treatments, significantly higher concentration of available phosphorus, Zinc and manganese was observed in T11 followed by T6. This implied that the rate of mineralization of added organic sources was enhanced causing a built up of available nutrients in the soil system. Further the data pointed out that statistically higher accumulation of DTPA extractable Cu in soil was perceived in T4 followed by T11 throughout the incubation period. The presence of organic matter may increase the availability of

Cu in soils owing to the formation of soluble complexing agents thereby decrease in the fixation of Cu in soils.

Further perusal of the result revealed that irrespective of different moisture regimes the amount of exchangeable NH_4^+ -N increased rapidly up to 30th day and then gradually declined till the end of the experiment. This rapid rise of NH_4^+-N content is attributable to the decomposition of the easily decomposable nitrogenous substances present in the organic materials. In general, soluble NO-3-N and amino acid N gradually increased throughout the incubation period which might be due to increased gross nitrification rates and conversion of other forms of N to amino acid form, respectively. Accumulation of hydrolysable organic NH_4^+ -N increased up to 60th day of incubation and thereafter, decreased at the end of experiment whereas a gradual decrease of total hydrolysable, unidentified hydrolysable and total N till the end of the experiment was observed under different moisture regimes. Irrespective of different treatments, significantly higher amount of exchangeable NH_4^+, hydrolysable NH_4^+, hexosamine, amino acid and total hydrolysable N was found at 80 per cent WHC followed by 60 per cent and 40 per cent WHC throughout the experiment. In contrast, 40 per cent WHC gave the significantly higher accumulation of non-hydrolysable and unidentified hydrolysable N followed by 60 per cent and 80 per cent WHC. This showed that higher moisture content encourage mineralization of non-hydrolysable and unidentified hydrolysable N. Significantly higher concentration of soluble NO_3^- and total N was observed at 80 per cent and 40 per cent WHC at the end of the experiment. Irrespective of moisture levels and incubation study, significantly higher amount of all the inorganic and organic forms of N as well as total hydrolysable and total N was recorded in N treated system over control throughout the incubation period. The magnitude of accumulation of exchangeable NH_4^+, hydrolysable NH_4^+, hexosamine and amino acid N was more in INM treated soil than sole addition of chemical fertilizer alone. Comparing among the different treatments, significantly higher amount of exchangeable NH_4^+ was observed in T4 followed by T11 and T6. However, similar effects of T4 and T11 was found in NO_3^-, hydrolysable NH_4^+, hexosamine, amino acid, total hydrolysable N and total N at the end of the experiment.

Conclusion

On the basis of the results obtained from the investigations carried out, it could be concluded that integration of different inorganic and organic sources of nitrogen markedly improved the organic carbon content, cation exchange capacity, available nutrient content, inorganic and organic fractions of nitrogen, nutrient uptake, plant height, yield attributes and yield of rice (straw and grain) and mustard (stover and seed) over single inorganic source application and untreated control. Nutrient release patterns as well as changes in nitrogen fractions were different between laboratory incubation under controlled condition and field experiment applied with the same treatments. Integrated nutrient management improves soil fertility while sustaining crop productivity.

REFERENCES

1. Serageldin, Ismail (2000). Dwanting research tasks. The Hindu Survey of Indian Agriculture, pp. 31.
2. Kumari, N. (2020). Foodgrains production in India: Trend and decompositions analysis. *Economic Affairs, 65*(3), 333-342.
3. Paroda, R.S. (1998). Meeting the food needs on a sustainable basis challenges and opportunities. (In) *Proc. Int. Symp. Rainfed rice for sustainable food security,* Cuttack, Orissa, India, pp. 1-15.
4. Virmani (1994). The New Challenges for Soils Research. *J. Ind. Soc. Soil Sci.,* 42(4): 516-524.
5. Singh, N.P., Sachan, R.S., Pandey, P.C. and Bisht, P.S. (1999). Effect of a decade long fertilizer and manure application on soil fertility and productivity of rice -wheat system in a Mollisol. *J. Ind. Sac. Soil Sci.,* 47(1): 72-79.
6. Bezbaruha, R., Sharma, R.C. and Banik, P. (2011). Effect of nutrient management and planting geometry on productivity of hybrid rice (*Oryza sativa* L.) cultivars. *Am J Plant Sci.,* 2: 297-302.
7. Varma, P. (2017). An Overview of Rice Economy. *Rice Productivity and Food Security in India,* 7-28.
8. Mathur, K.C. and Krishraiah, K. (1996). Perspective of rice research in India. *Indian Farming,* 32: 3-8.
9. Mittal, S. (2008) Demand Supply Trend and Projections of Food in India. *Ind. Council for Res. on International Economic Relations,* pp. 1-20.
10. Reddy, V. and Immanuelraj, K. (2017). Economic Affairs, All rights reserved Area, Production, Yield Trends and Pattern of Oilseeds Growth in India, June 2017 DOI: © 2017 New Delhi Publishers. Vol. 62, No. 2, pp. 327-334.
11. Jat, R. S., Singh, V. V., Sharma, P., and Rai, P. K. (2019). Oilseed brassica in India: Demand, supply, policy perspective and future potential. *OCL, 26,* 8.
12. Singh, J.K., Mishra, P.J. and Singh, R.A. (1998). Nutrient uptake, moisture use and economics of Indian mustard Brassica juncea (L.) Czernj and Cosson varieties at rainfed conditions. *Environ. Ecol.,* 16(1): 54-57.
13. Aulakh, M.S. and Pasricha, N.S. (1998). The effect of green manuring and fertilizer N application on enhancing crop productivity in mustard-rice rotation in semi-arid subtropical regions. Eur. *J. Agron.,* 8(1/2): 51-58.
14. Narang, R.S., Mahal, S.S. and Gill M.S. 1993. Effect of phosphorus and sulfur on growth and yield of toria. *Ind. J. Agron.* 38(4):593-597.
15. Devi, T. R., Devi, N. D., Vivekananda, Y., and Sharma, P. R. (2017). Genetic diversity analysis in Indian mustard (Brassica juncea L. Czern and Coss) genotypes using agro-morphological parameters. *Electronic Journal of Plant Breeding, 8*(3), 749-753.

16. Pillai, K.G. (1996). Current scenario in rice and prospects for sustainable rice production. *Fert. News*, 41 (2): 15-23.
17. Mishra, M.M. and Kapoor, K.K. (1992). Importance of chemical fertilizers in sustainable agriculture in India. *Fert. News*, 37: 47-52.
18. Abrol, I.P. and Gill, M. (1994). In: Regional Office for Asia and Pacific (FAO).
19. Jackson, M.L. (1973). Soil Chemical Analysis. Prentice Hall of India Pvt. Ltd., New Delhi
20. Subbiah B.V. and Asija, G.L. (1956). A rapid procedure for estimation of available Nitrogen in soils. *Curr. Sci.*, 25:259-260.
21. Lindsay,W.L. and Norvell,W.A. (1978).Development of DTPA-soil test for Zn,Fe,Mn and Cu.*Soil Sci. Soc.America J.*, 42: 421-428.
22. Elwell, W.T. and Gridley, J.A.F. (1967). *Atomic Absorption Spectrophotometry*, Pergamon Press Ltd. London.
23. Gomez, K.A. and Gomez, A.A. (1984). Statistical Procedure for Agricultural Research, Published by John Willey and Sons, New York: Pp. 630.

Chapter 4

Effect of Zinc on Soil Properties and Yield under Rice-Rapeseed Sequence

Sanjeev Yadav

ABSTRACT

Present investigation was conducted to assess the effect of zinc on nutrients uptake, yield and soil properties under rice-rapeseed cropping sequence at the Research Farm, College of Agriculture, CAU, Imphal. The experimental sitewas located at an elevation of 790 meter above mean sea level with 24°45' N latitude and 93°56' E longitude. The soil of the experimental field was clay in texture, high in carbon, medium in nitrogen, phosphorus, potassium, and Zn was above the critical level. Fifteen combinations of Zn treatments viz., T1: Control, T2: Zn @ 2.5kg/ha as $ZnSO_4.7\,H_2O$, T3: Zn @ 5 kg/ha $ZnSO_4.7\,H_2O$, T4: Zn @ 7.5 kg/ha $ZnSO_4.7\,H_2O$, T5: Zn @ 10 kg/ha $ZnSO_4.7\,H_2O$, T6: Zn @ 0.25 per cent foliar spray (2 times) as $ZnSO_4.7\,H_2O$, T7: Zn @ 0.5 per cent foliar spray (2 times) $ZnSO_4.7\,H_2O$, T8: Zn @ 2.5kg/ha + Zn @ 0.25 per cent foliar spray (2 time) as $ZnSO_4.7\,H_2O$, T9: Zn@ 2.5kg/ha +Zn 0.5 per cent foliar spray (2 times) as $ZnSO_4.7\,H_2O$, T10: Zn @ 5 kg/ha + Zn @ 0.25 per cent foliar spray (2 times) $ZnSO_4.7\,H_2O$, T11: Zn @ 5 kg/ha + Zn @ 0.5 per cent foliar (2 times) $ZnSO_4.7\,H_2O$, T12: Zn @ 7.5 kg/ha+ Zn @ 0.25 per cent foliar spray (2 times) $ZnSO_4.7\,H_2O$, T13: Zn @ 7.5 kg/ha +Zn @ 0.5 per cent foliar spray (2 times) as $ZnSO_4.7\,H_2O$, T14: Zn @ 10 kg/ha+ Zn @ 0.25 per cent foliar spray (2 times) $ZnSO_4.7\,H_2O$ and T15: Zn @ 10 kg/ha+ Zn @ 0.5 per cent foliar spray (2 times) $ZnSO_4.7\,H_2O$ were imposed to rice in three times replicated randomized block design (RBD). Rapeseed crop was grown with recommended doses of N, P_2O_5 and K_2O (40:20:20) and on residual Zn fertility. Results revealed that the application of Zn @ 7.5 kg/ha along with the foliar spray of Zn @ 0.5 per cent (two times: panicle initiation and booting stage) through $ZnSO_4.7\,H_2O$ recorded maximum values of growth and yield attributing characteristics of rice and resulted in significantly higher grain yield (5.77 t/ha) over control. However, application of Zn @ 10 *kg/ha along with the foliar spray of Zn @ 0.5 per cent as $ZnSO_4.7\,H_2O$ recorded the highest Zn content in grain of rice (27.9 mg/kg). With regards to the residual effect of Zn on rapeseed, residual effect of Zn @10 kg/ha along with the foliar application of Zn @ 0.5 per cent through Zn $ZnSO_4.7\,H_2O$ recorded the highest seed yield of rapeseed (692 kg/ha) and Zn content (85 mg/kg) and uptake (155 g/ha) under rice-rapeseed sequence. Zn application had failed to affect the soil properties after harvest of rice and rapeseed except for Zn status in soil. Increases in Zn level correspond to increase the soil available Zn, thus application of Zn @ 10 kg/ha along with the foliar application of Zn @ 0.5 per cent (two times) as $ZnSO_4.7\,H_2O$ in rice resulted in a maximum increment in the soil available Zn after harvest of each crop.*

Introduction

Zinc (Zn) is acrucial nutrient for optimum plant growth. Zn content in plants varies from 25 to 150 ppm. Inadequate or suboptimal Zn in soil causes Zn deficiency in plants. The plant shows Zn deficiencies symptoms when its concentration in plants drops below 20 ppm, and toxicities when the Zn concentration in leaf crosses the limit of400 ppm (Tisdale *et al.*, 1993). Plant uptakes Zn from soil mainly through the diffusion process. Zn2+ is the available form of Zn to the plants (Rashid, 2001). Zn is an essential element for many amino acids and protein synthesis in plant tissues. Zn deficiency in the plant system reduces the protein synthesis which shows that Zn is the main component of the ribosome (Klug and Rhodes, 1987). Pollen sterility in plants mainly causes due to poor Zn supply. Hence, Zn plays an important role in pollen tube formation (Pandey *et al.*, 2006). Inadequate Zn supply also negatively impacted various plant metabolic activities, and reduces overall plant growth and crop productivity (Pandey *et al.*, 2006).Sufficient Zn supply reduces the chances of oxidative damage in plant systems (Arvind and Prasad, 2005). Approximately 30 per cent of the global soils and more than two-billion-world population showed Zn deficiency (Cakmak *et al.*, 2008). Zn deficiency is widespread in soils of many Asian countries like Bangladesh, China, India, Pakistan, and Turkey (Prasad, 2005). In India around Forty-three percent soils are poor in Zn (Shukla *et al.*, 2014). Zn adsorption with different soil constituents, reliant on organic matter, pH pedogenic oxides, and redox potential is the main cause of Zn deficiency in soils (Havlin *et al.*, 2005).

Rice is the principal foods for >3.0 billion people in the globe. After wheat rice is the second main consumed cereal in the world, provides 15 per cent protein and 21 per cent energy necessities of humankind (Depar *et al.*, 2011). Globally rapeseed and mustard cultivated in 355.20 lakh ha area. India ranks first in the area and third in the production of rapeseed and mustard after China and Canada. In India, rapeseedmustard is grown during *Rabi* season across the country over an area of about 6.18 million ha. Food security in south Asia cannot be achieved without proper management of rice-upland systems (Yadav *et al.*, 2019 and Yadav *et al.*, 2000).

Generally, rice is grown in puddled submerged soil, where Zn is one of the most limiting nutrient elements. Inactive clay lattice bound Zn is the major Zn form in soil (>90 per cent of total Zn) (Mandal and Mandal, 1986). Exchangeable, water-soluble and, organic complexes, amorphous sesquioxide, and crystalline sesquioxide-bound types had very less contribution in total soil Zn (Mandal and Mandal, 1986). Waterlogging converted the native Zn in to the organic complex and amorphous bound types of Zn, consequently retarded the Zn availability to the plants. Deficiency symptoms of Zn in rice can be manifested as brown blotches and streaks on young leaves. Zn deficiency in plants increases the crop life cycle and reduces the yield. The critical Zn index in soil of sustainable rice production is 1.5 mg/kg (Yan, 2003). The average acquisition of Zn in rice for per tonne grain production is 50 g. The major portion of Zn (60 per cent) taken by rice plants remains in straw (IRRI, 2000). Zn deficiency is the major challenge for sustainable rice production in India (Chaudhary *et al.,* 2007).

Low economic yield along with the poor nutritional quality of plant products has been reported from plants grown on Zn deficient soils (Welch and Graham 1999). Cereal crops grown in poor Zn supplying soil had ~ 80 per cent less Zn content in grain (Cakmak *et al.,* 1997), Poor Zn content in grain reduce its bioavailability in human health system and causes Zn deficiency in mankind (Hussain *et al.,* 2012). In developing countries, anemia disease of human mainly causes mainly due to Zn deficiency where rice is the staple food (Biswas *et al.,* 2018). Undeniably, in area where cereal based products are the major food, Zn deficiency is the major challenge for providing the healthy diet to the public (Welch, 1993). A diet should have 15 mg Zn/day for healthy human adults (Hafeez *et al.,* 2013). Moreover, the sustainability of rice centered sequences is threatened by Zn deficiency. Zinc sulphate is the novel product generally use to remediation of Zn deficiency. Transplanted rice responded well to zinc fertilization (Mollah *et al.,* 2009 and Fageria *et al.,* 2011). Zn played crucial role in plant metabolism like increased in carbohydrate synthesis and its transportation to the grain production site and resulted in high crop yield (Sachdev *et al.,* 1988 and Pedda Babu *et al.,* 2007). Combined application (soil and foliar) Zn fertilizers are effective strategy to mitigate Zn stress. More Zn dense grains can be obtained by the constant supply of Zn in the rhizosphere (Jena *et al.,* 2006). Soilfertilization of Zn increases crop yield about 5 per cent, however, foliar application of Zn can also increaseZn density in grain (Wissuwa *et al.,* 2008). However, Torun *et al.* (2001) reported that the micronutrients stress in deeper soilcan also be corrected by foliar spray. Keram *et al.* (2012) stated that the Zn fertilization does not affect the soil properties like pH, EC, SOC, CEC, and calcium carbonate content, but increases the soil Zn status considerably.

Number of branches per pant, siliqua per plant and seed per siliqua are major yield governing parameters in in rapeseed and mustard (Diepenbrock, 2000 and Riley *et al.,* 2000). Zn fertilization significantlyenhances the seeds per siliqua and siliqua/plant in rapeseed and mustard and ultimately lead to high seed yield (Riley *et al.,* 2000). Zn spray at stem elongation and flowering stages enhanced seed yield (by 11 per cent and 17 per cent, respectively), protein and oil content (Omidian *et al.,* 2012). Balance fertilization is a desirable characteristic of sustainable cropping systems. Nutrients applied to the preceding crops exhibit residual effects on succeeding crops (Babu *et al.*, 2013) particularly in acidic soil; hence fertilization must be done keeping the whole cropping system in view rather than the individual crops. In rice-wheat system, zinc fertilization imposed to rice had considerable residual effect on succeeding wheat(Amanullah *et al.*, 2016). Balance fertilization is critical for realizing higher crop yield but crops are generally supplied with major nutrients and the application of micronutrients is largely ignored by the farmers. Evidence supported that the application of Zn increases the productivity of rice and rapeseed. However, very little information is available about the direct and residual effect Zn on the crop yield and soil properties of rice-rapeseed cropping systems. The literature is also lacking in a comparative assessment of foliar and soil application of zinc in various combinations on soil properties and productivity of rice- rapeseed system in acidic soils. Keeping in view the present investigation entitled "*Effect of Zinc on Soil Properties and Yield under Rice-Rapeseed Sequence*" was conducted during 2019-20 to observe the effects of Zn on soil properties and yield of rice-rapeseed system with following objectives:

1. To study the effect of zinc on nutrient uptake and yield of ricerapeseed cropping sequence
2. To study the effect of zinc on soil properties in a rice-rapeseed cropping sequence

Materials and Methods

Description of Study Site

The experiment was conducted at the Research Farm, College of Agriculture, Central Agricultural University, Imphal, situated at 24° 48' 43 N and 93° 53' 29" E and an altitude of 770 m above the mean sea level. The climate of the experimental site (Imphal) was humid subtropical with dry winters and a wet and hot monsoon season. The region is a witness of the large variation in weather both at spatial and temporal scales. South-west monsoon plays an important role in contributing to a huge amount of rainfall. The Imphal valley receives about 1,320 mm of rain annually. The region has a large variation in temperature across the year. January is the coldest month while May is the hottest month. The minimum and maximum temperatures during the cropping year were 7.5 and 30.7°C, respectively. Total rainfall of 1168.3 mm was received during the experimental period.

Experimental Treatments and Design

The experiment was laid out in three times replicated Completely Randomized Block Design (CRBD). Treatment details are given below:

Treatments (for rice)

T1: Control

T2: Zn @ 2.5 kg/ha as a $ZnSO_4.7\ H_2O$

T3: Zn @ 5.0 kg/ha as a $ZnSO_4.7\ H_2O$

T4: Zn @ 7.5 kg/haas a $ZnSO_4.7\ H_2O$

T5: Zn @ 10.0 kg/ha as a $ZnSO_4.7\ H_2O$

T6: Zn @ 0.25 per cent foliar spray (2 times) as a $ZnSO_4.7\ H_2O$

T7: Zn @ 0.5 per cent foliar spray (2 times) as a $ZnSO_4.7\ H_2O$

T8: Zn @ 2.5 kg/ha + Zn @ 0.25 per cent foliar spray (2 time) as a $ZnSO_4.7\ H_2O$

T9: Zn @ 2.5 kg/ha +Zn 0.5 per cent foliar spray (2 times) as a $ZnSO_4.7\ H_2O$

T10: Zn @ 5.0 kg/ha + Zn @ 0.25 per cent foliar spray (2 times) as a $ZnSO_4.7\ H_2O$

T11: Zn @ 5.0 kg/ha + Zn @ 0.5 per cent foliar (2 times) as a $ZnSO_4.7\ H_2O$

T12: Zn @ 7.5 kg/ha+ Zn @ 0.25 per cent foliar spray (2 times) as a $ZnSO_4.7\ H_2O$

T13: Zn @ 7.5 kg/ha +Zn @ 0.5 per cent foliar spray (2 times) as a $ZnSO_4.7\ H_2O$

T14: Zn @ 10.0 kg/ha+ Zn @ 0.25 per cent foliar spray (2 times) as a $ZnSO_4.7\ H_2O$

T15: Zn @ 10.0 kg/ha+ Zn @ 0.5 per cent foliar spray (2 times) as a $ZnSO_4.7\ H_2O$

Note: - A recommended dose of N, P, and K will be applied in all the treatments.

Details of the Experiment

Experimental Design: Randomized Block Design (RBD)

No. of replications: 3

No of treatments: 15

Test crop: Rice (CAU-R1) and Rapeseed (M-27)

Total no. of plots: 45

Plot size: 5 m x 3 m

The recommended dose of fertilizer: Rice 60:40:30 N $P_2O_5K_2O$ kg/ha

Rapeseed 40:20:20 N $P_2O_5K_2O$ kg/ha

Date of nursery sowing of rice: 16th June, 2019

Date of transplanting of rice: 17th July, 2019

Date of harvesting of rice: 8th November, 2019

Date of sowing of rapeseed: 2nd December, 2019

Date of harvesting of rapeseed: 10th March, 2020

Rapeseed crop was grown with the recommended dose of fertilizers to assess the residual effect of Zn. Rice variety CAU R1 and Rapeseed variety M27 was used during the experiment. Recommended agronomic practices for rice and rapeseed were adopted for healthy growth of the crop in field trial.

Soil Sampling and Plant Analysis

The soil of the Research Farm, College of Agriculture, Central Agricultural University, Imphal is acidic in reaction. Before the setup of the experiment, soil samples from the experimental field were collected randomly from 0–15 cm soil depth from several spots. A composite sample was made for the soil analysis. The soil of the experimental site was silty clay and having good water holding capacity. The data on soil properties before the start of the study of the experimental field is presented in Table 4.1.

Table 4.1. Initial Physio-chemical Properties of Experimental Field

Soil properties	*Results*
Textural Class	Clayey
Sand (per cent)	6.14
Silt (per cent)	8.9
Clay (per cent)	82.25
pH (1:2.5 soil: water ratio)	5.09
EC (dS m-1)	0.20
CEC (Cmol (p+)/kg)	17.01
Organic carbon (per cent)	1.09
Available N (kg/ha)	255.02
Available P_2O_5 (kg/ha)	28.17
Available K_2O (kg/ha)	239.98
DTPA Zinc (ppm)	0.70

Before the start of the experiment, soil samples were collected from five points in each replication using a soil auger from the depth of 0–15 and made a composite soil sample. The composite soil samples were analyzed for mechanical composition, soil organic carbon, soil pH, EC, CEC, available-N, available phosphorus, available potassium, and DTPA-extractable zinc. Plot-wise soil samples were collected after harvest of rice and rapeseed from a depth of 0–15 cm and analyzed for pH, EC, CEC, available-N, available phosphorus, available potassium, DTPA-extractable zinc. The collected bulk soil samples before the conductance of the experiment and after harvest of each crop were air-dried at room temperature, clods were broken by giving gentle strokes with a hammer, and the foreign material was removed, and sieved with 2 mm sieve. The processed soil samples were stored in airtight plastic bags until the pending analysis. Following methods were used for soil and plant analysis:

Soil/Plant Analysis	*Methods of Analysis*
Mechanical Analysis	Bouyoucos, 1962
Maximum Water Holding Capacity	Saha, 2008
Soil pH	Jackson, 1973

Soil/Plant Analysis	*Methods of Analysis*
Electrical Conductivity	Jackson, 1973
Cation Exchange Capacity	Borah *et al.* (1987).
Organic Carbon	Jackson, 1973
Available Nitrogen	Subbiah and Asija, 1956
Available Phosphorus	Jackson, 1973
Available Potassium	Jackson, 1973
Available Zinc	Lindsay and Norvel (1978)

Statistical Analysis

The data related to each parameter were analyzed as per the procedure of analysis of variance (ANOVA) and significance of Completely Randomized Block Design (CRBD) tested by the 'F' test (Gomez and Gomez, 1984). Standard Error of Means (SEm±) and Least Significant Difference (LSD) at 5 per cent level of significance were worked out for each parameter.

Brief Findings

In this field study Fifteen treatment combinations of Zn *viz.,* T1: Control (no Zn), T2: Zn @ 2.5 kg/ha as $ZnSO_4.7\ H_2O$, T3: Zn @ 5 kg/ha as $ZnSO_4.7\ H_2O$, T4: Zn @ 7.5 kg/ha as $ZnSO_4.7\ H_2O$, T5: Zn @ 10 kg/ha as $ZnSO_4.7\ H_2O$, T6: Zn @ 0.25 per cent foliar spray (2 times) as $ZnSO_4.7\ H_2O$, T7: Zn @ 0.5 per cent foliar spray (2 times) as $ZnSO_4.7\ H_2O$, T8: Zn @ 2.5 kg/ha + Zn @ 0.25 per cent foliar spray (2 time) as $ZnSO_4.7\ H_2O$, T9: Zn @ 2.5 kg/ha + Zn 0.5 per cent foliar spray (2 times) as $ZnSO_4.7\ H_2O$, T10: Zn @ 5 kg/ha + Zn @ 0.25 per cent foliar spray (2 times) as $ZnSO_4.7\ H_2O$, T11: Zn @ 5 kg/ha + Zn @ 0.5 per cent foliar (2 times) as $ZnSO_4.7\ H_2O$, T12: Zn @ 7.5 kg/ha+ Zn @ 0.25 per cent foliar spray (2 times) as $ZnSO_4.7\ H_2O$, T13: Zn @ 7.5 kg/ha +Zn @ 0.5 per cent foliar spray (2 times) as $ZnSO_4.7\ H_2O$, T14: Zn @ 10 kg/ha+ Zn @ 0.25 per cent foliar spray (2 times) as $ZnSO_4.7\ H_2O$ and T15: Zn @ 10 kg/ha+ Zn @ 0.5 per cent foliar spray (2 times) as $ZnSO_4.7\ H_2O$ were imposed to rice crops in three times replicated Randomized Block Design (RBD). Zn under foliar application treatments was applied at panicle initiation and booting stages of rice. Rapeseed crop was grown with full recommended doses of $N.P_2O_5$ and K_2O and on residual Zn fertility levels. The salient findings emerging from, this experimentation are briefly summarized below.

Effect of Zn Application on Rice Growth, and Yield, and Zn Content and Uptake in Rice

1. Zn nutritional treatments increased the plant height of rice over the control. Among the Zn treatments, T13 had significantly higher rice plant height than those of T1 and T6 and remained at par with the rest of the treatment in terms of plant height of rice.

2. Soil application of 7.5 to 10 kg Zn through $ZnSO_4.7\ H_2O$ produced significantly higher productive tillers per hills as compared to control treatment. Among all the treatments, T13 produced the maximum productive tillers per hill.
3. Rice plants grew with soil application of 5 to 7.5 kg/ha produced significantly higher panicle length as compared to that of control treatment. Treatment T13produced the longest panicle while the control treatment plant bearded the shortest panicle.
4. Soil application of 7.5 and 10 kg Zn/ha alone with a foliar spray of 0.25 and 0.5 per cent Zn at panicle initiation and booting stage enhances the grains per panicle over soil application of Zn alone (2.5 to 10 kg/ha). Among the treatments, T13 recorded the maximum grains per panicle which were at with T14, T15, T12, and T5.
5. Zn application failed to affect the test weight of rice (1,000 grain weight) of rice. However, among the treatments, the maximum 1,000 grain weight was observed under T13 while the lowest test weight was observed under control treatment.
6. Soil application of 5 to 10 kg Zn/ha through $ZnSO_4.7\ H_2O$ increased the rice grain yield by 16.6 to 26.7 per cent over control. Combined application of Zn increased grain yield of rice up to when applied as 7.5 kg Zn/ha in soil along with two foliar sprays of 0.5 per cent Zn at panicle initiation and booting stage. Among all the treatments, T13 produced the highest grain yield of rice. With regard to the straw yield, the combined application of Zn (soil + foliar), T12 to T15 produced significantly higher rice straw yield (7.25 to 7.40 t/ha) than that of control (5.45 t/ha).
7. Among the Zn nutrition treatments, T13 produced the maximum biological yield of rice which was significantly higher than those under T1, T2, T6, T7, and T8 treatments and remained statistically at par with the rest of the treatments. However, the harvest index of rice was not significantly affected Zn nutrition.
8. All the Zn nutrition treatments increased the Zn content in rice grain. However, among all the treatments, T15 recorded the maximum Zn content in rice grain.
9. Among the Zn treatments, application of Zn @10 kg/ha through zinc sulphate had the maximum Zn content in rice straw. When comparing the foliar applied Zn treatments, foliar application of 0.5 per cent Zn through zinc sulphate loaded higher Zn in rice straw than 0.25 per cent foliar application of Zn at panicle initiation and booting stages of rice. Content of Zn in rice straw (80.1 to 88.2 g/kg DM) increased further over soil and foliar applied treatments when soil application of Zn combined with foliar application of Zn.

10. Among the treatments, basal application of Zn @10 kg/ha + foliar application 0.5 per cent Zn at panicle initiation and booting stage observed statistically significant higher uptake of Zn in rice grain than those in rice grains of rest of the treatments.
11. Zn fertilization enhanced Zn uptake in rice straw over that of control and varied among the Zn applied treatments. Maximum Zn uptake in rice straw was found in T15. Zn uptake in rice straw under T15 was at par with T4, T5, and T9 to T14 and significantly higher than those of the other treatments.
12. Application of Zn @ 10 kg/ha along with the foliar spray of Zn @ 0.5 per cent at two times (panicle initiation and booting stage) (T15) recoded the maximum total Zn uptake in rice.

The Residual Effect of Zn on Rapeseed Growth and Yield, and Zn Content and Uptake

1. Among the treatments, treatment T15 had the tallest plant of rapeseed. However, among the soil applied treatments, T5 had significantly greater plant height than those under other soil applied treatment, control, and foliar treatments. Foliar application of Zn in rice did not have any residual effect on plant height of the following rapeseed.
2. The residual of Zn applied in rice on rapeseed was more pronounced with higher doses of Zn than lower doses. Residual effect of Zn @ 5 kg/ha and more significantly enhanced the branching ability of rapeseed. However, foliar application of Zn on rice did show any residual effect on the number of branches of rapeseed. Treatments T3 to T5 and T10 to T15 had a significantly higher number of branches than that of control treatment.
3. Zn applied through $ZnSO_4.7\ H_2O$ in rice had a considerable residual effect on the silique per plant of rapeseed. Among the Zn nutritional treatments, the residual effect of T15 had a significantly higher number of silique (24.8) per plant of rapeseed than those under T1, T2, T6, T7, T8, and T9 and remained statistically at part with the rest of the Treatments.
4. The seeds per silique in succeeding rapeseed was varied significantly due to the residual effect of Zn. T15 had a significantly higher number of seeds (14.4) per silique than those under T1, T2, T6, T7, T8, and T9 and remained statistically at part with the rest of the Treatments.
5. 1,000 seed weight of rapeseed was significantly affected by residual of Zn applied in rice. T15 had a significantly higher 1,000 seed weight (2.74 g) of rapeseed over the control.
6. The residual effect of among the soil applied treatments, T5 (682 kg/ha) produced significantly higher seed yield of rapeseed than those of T6, T7, and control treatment and remain at par with rest of the treatments.

However, among the combined applied (soil + foliar) Zn treatments, T15 (692 kg/ha) produced significantly greater seed yield of rapeseed as compared to T1, T2, T6, and T7, and remained at par with all other treatments.

7. The residual effect of 10 kg soil application of Zn/ha as a $ZnSO_4.7\ H_2O$ increased the stover yield of rapeseed by 22.5 per cent over control treatment. However, the residual effect of soil application 10 kg Zn/ha along with a foliar spray of 0.5 per cent at panicle initiation and booting stages of rice (T15) increased the stover yield of rapeseed by 23.0 per cent over control.
8. The residual effect of Zn @ 10 kg/ha increased the biological yield of rapeseed by 18.6 per cent over control treatment. The residual effect of soil application 10 kg Zn/ha along with a foliar spray of 0.5 per cent at panicle initiation and booting stages of rice (T15) increased the biological yield of rapeseed by 19.5 per cent over control. With regards to the harvest index of rapeseed, the residual effect of Zn had failed to affect the HI of rapeseed significantly.
9. Among the Zn treatments imposed to preceding rice crop, the residual effect of soil application of Zn @ 10 kg/ha along with a foliar spray of Zn @ 0.5 per cent at panicle initiation and booting stages of rice (T15) increased the Zn content in the seed of rapeseed by 17.6 per cent and in stover by 38.3 per cent over control.
10. The residual effect of Zn @ 10 kg/ha enhances the Zn uptake in the seed of rapeseed by 26.3 per cent over control treatment (44.9 g/ha). In the case of combined applied treatments, the residual effect of soil application 10 kg Zn/ha along with a foliar spray of 0.5 per cent at panicle initiation and booting stages of rice (T15) recorded significantly higher Zn uptake in the seed of rapeseed compare to all other treatments except T12 to T14 and T4 to T5.
11. The residual effect of Zn @ 10 kg/ha increased the Zn uptake in stover of rapeseed by 57.2 per cent over control. In the case of combined applied treatments, the residual effect of soil application 10 kg Zn/ha along with a foliar spray of 0.5 per cent at panicle initiation and booting stages of rice (T15) recorded significantly higher Zn uptake in stover of rapeseed over control. The residual effect of soil application 10 kg Zn/ha along with a foliar spray of 0.5 per cent at panicle initiation and booting stages of rice (T15) increased the Zn uptake in stover of rapeseed by 70.3 per cent over control.
12. Among the Zn nutrition treatments, the residual effect of Zn @ 10 kg/ha along with a foliar spray of 0.5 per cent at panicle initiation and booting stages of rice (T15) recorded significantly higher total Zn uptake in rapeseed compare to all other treatments except T12 to T14 and T4 to T5.

The residual effect of Zn @ 10 kg/ha along with a foliar spray of 0.5 per cent at panicle initiation and booting stages of rice (T15) increased the total Zn uptake in rapeseed by 52.5 per cent over without Zn treatment.

Effect of Zinc Application on Soil Properties in a Rice-Rapeseed Cropping Sequence

Soil Properties after Harvest of Rice

1. Application of Zn did not change pH of soil significantly after the harvest of rice, however, among the treatments, soil pH was the maximum under T2 and T9 while the minimum was under T5 and T14.
2. Soil EC after the harvest of rice did not affect by Zn nutrition. However, soil EC was ranged from 0.19 to 0.22 dS/m. Treatment T5, T8, T13, and T15 had the lowest EC while T1, T2, T6, and T7 had the maximum soil EC after the harvest of rice.
3. After the harvest of rice soil Cation exchange capacity varied from 18.2 to 21.6 Cmol (P+)/kg across the treatments. However, variations in CEC among the treatments were not found significant with the application Zn after the harvest of rice.
4. Application of Zn had failed to affect the soil organic carbon content after harvest of rice. However, soil organic carbon content increased with increasing the doses of Zn application. Among the treatments, T15 had the maximum content of soil organic carbon while T1, T7, and T8 had the lowest content of soil organic carbon.
5. Available nitrogen after the harvest of rice was not affected by Zn application. However, the highest content of available nitrogen was found under T15 (286.7 kg N/ha) and the lowest was under T7 (265.3 kg N/ha).
6. Application of Zn did not affect available phosphorus significantly after the harvest of rice. However, successive increments in Zn doses decrease the soil P status and T15 had the lowest available phosphorus (11.8 kg/ha) after the harvest of rice.
7. Soil available potassium ranged from 209.7 to 231.7 kg/ha across the treatments after the harvest of rice. Despite the non-significant effect of Zn nutrition on soil K status, the available potassium increased with increasing the levels of Zn application.
8. Soil available Zn was significantly higher in treatments that received Zn either through soil application or combined application than those in soils under control and foliar applied treatments. Among the treatments, T14 and T15 had the maximum available soil Zn while the lowest available Zn was under T6.

Soil Properties after Harvest of Rapeseed

1. The residual effect of applied Zn did not affect soil pH after harvest of rapeseed in rice-rapeseed sequence. However, soil pH was the maximum under T2, T4, and T7 while the minimum was under T11 and T14 after harvest of rapeseed.
2. The residual effect of Zn had failed to affect the soil EC. Although, treatment T11 had the lowest EC while T14 had the maximum soil EC after harvest of rapeseed.
3. Variations in CEC among the treatments were not found significant with the residual effect of Zn application after harvest of rapeseed. Although treatment T15 had the maximum CEC and T1 had the lowest CEC after harvest of rapeseed.
4. The residual effect of Zn application had failed to affect the soil organic carbon content after harvest of rapeseed. However, among the treatments, T15 had the maximum content of soil organic carbon while T1 and T7 had the lowest content of SOC after harvest of rapeseed.
5. Residual effect of Zn application had failed to affect the available nitrogen in soil after harvesting of rapeseed. The highest content of available nitrogen was found under T15 (293.3 kg N/ha) and the lowest content of available nitrogen was under T1 (275.7 kg N/ha).
6. Among the Zn treatments, T15 had the lowest available phosphorus (14.1 kg/ha) after the harvest of rapeseed. Although differences among the treatments were statistically non-significant.
7. Soil available potassium ranged from 210.7 to 226.0 kg/ha across the treatments after harvesting of rapeseed. Soil available potassium was higher with residual effect combined application of Zn (soil + foliar application) after harvest of rapeseed than those of soil and foliar application alone and control.
8. Soil available Zn in soil applied treatments ranged from 0.79 to 1.05 mg/kg soil after harvest of rapeseed. Among the treatments, T14 (1.07 mg/kg soil) and T15 (1.09 mg/kg soil) had the maximum available soil Zn while the lowest available Zn was under T1 (0.68 mg/kg soil) after harvest of rapeseed.

Conclusions

1. Soil application of Zn @ 7.5 kg/ha + Zn @ 0.5 per cent foliar spray (2 times) through $ZnSO_4.7\ H_2O$ at panicle initiation and booting stages produced significantly higher rice yield than other treatments.
2. Application of Zn @ 10 kg/ha along with the foliar spray of Zn @ 0.5 per cent (2 times) through $ZnSO_4.7\ H_2O$ resulted in the maximum Zn content

in both grain and straw of rice followed by the application of Zn @ 10 kg/ ha as a basal and foliar spray of Zn @ 0.25 per cent Zn as a $ZnSO_4.7\ H_2O$.

3. Application of Zn @ 10 kg/ha along with the foliar application of Zn @ 0.5 per cent through $ZnSO_4.7\ H_2O$ in rice exerted the maximum residual effect on yield of rapeseed under rice-rapeseed sequence followed by the application of Zn @ 10 kg/ha along with foliar spray of Zn @ 0.25 per cent in rice.
4. Neither soil application nor the foliar and combined application of Zn affected the soil properties after harvest of each crop except in available Zn. Increases in Zn level correspond to the increase in the soil available Zn, thus l application of Zn @ 10 kg/ha along with the foliar application of 0.5 per cent Zn (two times) through $ZnSO_4.7\ H_2O$ in rice resulted in the maximum increment in the soil available Zn.

REFERENCES

1. Tisdale, S.L., Nelson, W.L. and Beaten, J.D. (1984). Zinc In soil Fertility and Fertilizers.Fourth edition, Macmillan Publishing Company, New York.382-391.
2. Rashid, A. (2001). Secondary and micronutrients; Soil Science. National Book Foundation, Islamabad, p. 372-379.
3. Klug, A. and Rhodes, D. (1987). ¯Zinc fingers : A noval protein motif for nucleic acid recognition. *Trends Biochem. Sci.*, 12: 464-469
4. Pandey, N., Pathak, G.C. and Sharma, C.P. (2006). Zinc is critically required for pollen function and fertilization in lentil. *J. Trace Elem. Med. Bio.,* 20: 89-96.
5. Arvind, P and Prasad, MN (2005). Modulation of cadmium-induced oxidative stress in *Ceratophyllumdemersum* by zinc involves ascorbate glutathione cycle and glutathione metabolism. *Plant Physiol. Biochem.,* 43:107-116
6. Cakmak, I., Kalayni, M., Ekiz, H, Braun, H.J., Kilinc, Y. and Yilmaz, A. (2008). Micronutrient deficiencies in global crop production. *Field Crops Res.,* 60-77.
7. Prasad, R. (2005). Rice wheat cropping systems. *Adv. Agron.,* 86: 255-339.
8. Shukla, A.K., Tiwari, P.K. and Prakash, C. (2014). Micronutrient deficiencies *vis* a *vis*food and nutritional security of India. *Indian J. fert.,* 10(12): 94-112.
9. Havlin, J.L., Beaton JD, Tisdale, SL and Melson, WL (2005) Soil fertility and fertilizer: An introduction to nutrient management 7th edition, Pearson education, Inc, Upper So ddle River, New Jersey.
10. Depar, N., Rajpar, I., Memon, M.Y., Imtiaz, M. and Zia-ulhassan, (2011). Mineral nutrient densities in some domestic and exotic rice genotypes. *Pak. J. Agri. Agril. Engg. Vet. Sci.,* 27: 134–142.
11. Yadav, G.S., Das, A., Lal, R., Babu, S., Datta, M., Meena, R.S., Patil, S.B., Singh, R. (2019). Impact of no-till and mulching on soil carbon sequestration under rice (*Oryza sativa* L.)- rapeseed (*Brassica campestris* L. var. rapeseed) cropping

system in hilly agroecosystem of the Eastern Himalayas, Indian *Agric. Ecosyst. Environ.,* 275: 81-92.

12. Yadav, R.L., Dwivedi, B.S., Prasad, K., Tomar, O.K., Shurpali, N.J. and Pandey, P.S. (2000). Yield trends, and changes in soil Organic-C and available NPK in a long-term ricewheat system under integrated use of manures and fertilizers. *Field Crops Res.,* 68: 219-246.
13. Mandal, L.N. and Mandal, B. (1986). Zinc fractions in soils in relation to zinc nutrition of lowland rice.*Soil Sci.,* 142(3): 141-148
14. Yan, X.X. (2003). Research on genotypic difference in zinc nutrition of rice. *Guangdang Trace Element Sci.,* 10: 6-9
15. IRRI. (2000). Nutritional disorders and nutrient management in rice. Int. Rice Res. Ins. Manila, Philippines.
16. Chaudhary, S.K. and Sinha, N.K. (2007). Effect of levels of nitrogen and zinc on grain yield and their uptake in transplanted rice.*Oryza,* 44(1): 44-47.
17. Welch, R.M. and Graham, R.D. (1999). A new paradigm for world agriculture: meeting human needs. Productive, sustainable, nutritious. *Field Crops Res.,* 60: 1-10.
18. Cakmak, I., Ekiz, H., Yilmaz, A., Torun, B., Koleli, N., Gultekin, I., Alkan, A. and Eker, S. (1997).
19. Differential response of rye, triticale, bread and durum wheat to zinc deficiency in calcareous soils.*Plant Soil,* 188: 1-10
20. Hussain, S., Maqsood, M.A., Rengel, Z. and Aziz, T. (2012). Biofortification and estimated human bioavailability of zinc in wheat grains as influenced by methods of zinc application. *Plant Soil,* doi:10.1007/s11104-012-1217-4
21. Biswas, J.C., Haque, M.M., Khan, F.H., Islam, M.R., Dipti, S.S., Akter, M. and Ahmed, H.U. (2018). Zinc fortification: Effect of polishing on parboiled and unparboiled rice.*Curr. Plant Bio.,* 16: 22–26
22. Hafeez, B, Khanif, YM and Saleem M (2013) Role of zinc in plant nutrition -A review. *Am. J. Exp. Agric.,* 3(2): 374-391.
23. Mollah, M.Z.I., Talukder, N.M., Islam, M.N. and Ferdous, Z. (2009). Effect of nutrients content in rice as influenced by zinc fertilization.*World Applied Sci. J.,* 6(8): 1082-1088.
24. Fageria, K.N., Santos dos, BA and Cobucci, T (2011) Zinc nutrition of lowland rice.*Comm. Soil Sci. Plant Anal.,* 42(14): 1719-1727
25. Sachdev, P., Dep, D.L. and Rastogi, D.K. (1988). Effect of varying levels of zinc and manganese on dry matter yield and mineral composition of wheat plant at maturity. *J. Nuclear Agric. Biol.,* 17: 137- 143.
26. PedaBabu, P., Shanti, M., Rajendra Prasad, B. and Minhas, P.S. (2007). Effect of zinc on rice in rice –black gram cropping system in saline soils.

27. Jena, P.K., Rao, C.P. and Subbiah, G. (2006). Effect of zinc management practices on growth, yield and economics in transplanted rice. *Oryza,* 43(4): 326-328
28. Wissuwa, M., Ismail, A. and Graham, R. (2008). Rice grain zinc concentrations as affected by genotype, native soil zinc availability, and zinc fertilization. *Plant Soil.,* 306: 37-48.
29. Keram, K.S., Sharma, B.L. and Sawarkar, S.D. (2012). Impact of Zn application on yield, quality, nutrients uptake and soil fertility in a medium deep black soil (Vertisols). *Int. J. Environ. Sci. Technol.,* 1 (5): 563 -571.
30. Diepenbrock, W. (2000). Yield analysis of winter oilseed rape (*Brassica napus* L.) a review. *Field Crops Res.* 67: 35-49
31. Riley, T.G., Zhao, F. and McGrath, S.P. (2000). Available of different form of sulphur fertilizer on wheat and oil seed rape. *Plant Soil,* 222: 139- 147
32. Omidian, A., Seyadat, S.A., Naseri, R. and Moradi, M. (2012). Effects of zinc-sulfate foliar on yield, oil content and seed protein of four cultivars of canola. *Iranian J. Agric. Sci.,* 14: 26-28.
33. Babu, Subhash, Rana, D. S, Prasad, D. and Pal, S. (2013). Effect of sunflower stover and nutrients management on productivity, nutrient economy and phosphorus use efficiencies of pigeonpea (*Cajanuscajan*)-sunflower (*Helianthus annuus*) cropping system. *Indian J. Agric. Sci.,* 83(2): 203-9
34. Amanullah and Inamullah (2016). Residual phosphorus and zinc influence wheat productivity under rice–wheat cropping system. *Springer Plus* 5:255 DOI 10.1186/s40064-016- 1907-0
35. Bouyoucos, G.H. (1951). A recalibration of the hydrometer for making mechanical analysis of soils. *Agron. J.* (43): 434-438
36. Saha, P.K., Ishaqueb, M., Salequeb, M.A., Miahb, M.A.M., Panaullahb, G.M. and Bhuiyanb N.I.(2007). Long term integrated nutrient management for rice based cropping pattern: effect on growth, yield, nutrient uptake, nutrient balance sheet, and soil fertility. *Commun. Soil Sci.Plant Analysis,* 38(5-6): 579-610.
37. Jackson, M.L. (1973). Soil Chemical Analysis. Prentice Hall of India Pvt. Ltd., New Delhi, 498.
38. Borah, DK, Bordoloi, PK, Karmakar, RM, Baruah, NG and Das, M (1987) Practical Manual of Fundamental of Soil Science (Part-III), Jorhat, Assam
39. Subbiah, B.V. and Asija, G.L. (1956). A rapid procedure for estimation of available N in soils. *Curr. Sci.,* 25: 259-260.
40. Lindsay, W.L. and Norvell, W.A. (1978). Development of DTPA soil test for zinc, manganese and copper. *Soil Sci. Soc. Am. J.,* 42: 421-428.
41. Gomez, K.A. and Gomez, A.A. (1984). Statistical procedures for Agriculture Research. 2nd Edition, a Wiley International Science Publication, New York, Pp. 324.

Section C

Nutrient Management in Legumes

Chapter 5

Comparative Efficiency of Zinc with and without Organic Manure in Growth and Effect of Organic Amendments in Zinc Nutrition of Chickpea (*Cicer arietinum*)

T. Vikas kumar

ABSTRACT

A study was conducted in Churachandpur block, Churachandpur district of Manipur with the objective of to study zinc distribution in Jhum soils of Churachandpur district and the comparative efficiency of organic and inorganic sources in affecting the growth, yield and uptake of zinc by chickpea and to estimate the critical limit of zinc in soils and crop. To achieve the objectives of this study was carried out in pots under net house during Rabi season November 2016-17 at the research farm of College of Agriculture, CAU, Iroisemba, Imphal. Chickpea (var GJG0891) was grown as test crop in pots with different treatment combination of zinc with and without vermicompost. The results revealed that, the mean distribution of zinc in surface soils of different villages of Churachandpur district recorded is 2.28 mg kg^{-1} *which is deficient compared to critical limits in soils as 2.60 mg kg*$^{-1}$*. The comparative efficiency of zinc and zinc with vermicompost showing significant increase in chickpea plant height (39.1 cm) and number of branches (3.00) per plant were recorded with increased level of zinc with vemicompost T8 (Zn 10 kg ha*$^{-1}$*+ 5 t ha*$^{-1}$ *vermicompost) and decent dry matter weight*

(11.87 g pot^{-1}) and uptake of zinc (162.68-289.17μg pot^{-1}) were recorded with increased level of zinc with vemicompost T8 (Zn 10 kg ha^{-1} + 5 tha^{-1} vermicompost) in different villages of Churachandpur having Jhum soils. In Churachandpur block, soil physico chemical parameter, zinc critical limits in soil (2.60 mg/kg^{-1}) and chickpea crop (12.20 mg kg^{-1}), growth and dry matter yield with zinc and vermicompst (Zinc 10kgha^{-1}+ 5tha^{-1} vermicompost) were found.

Introduction

Chickpea (*Cicer arietinum* L.) is one of the important crops of the world and is consumed by people across the globe. It originated from Western Asia (Turkey). For vegetable purpose, it has become one of the most popular crops in the country. Madhya Pradesh ranks first in Area (2.6 Mha), production (2.4 m.t) and productivity of 930 kg/ha followed by Rajasthan among the Indian states. The three states Madhya Pradesh, Rajasthan, Uttar Pradesh account for 84 per cent of Area and 86 per cent production in the country. National average productivity is 810 kg/ha whereas higher average productivity is 1.8 t/ha in Egypt. Chickpea grows on wide range of soils from medium to heavy black soils. It does well on Black cotton soils and sandy loams. The crop needs an optimum pH of 6.0 to 7.5 (>8.5pH not suitable). Chickpeas is highly susceptible to damage by water -logging, saline and alkaline conditionsThis *Rabi* pulse crop requires cool humid weather and is mainly suitable to North India. It is suited for moderate rainfall areas of 400-700mm. Waterlogging results into wilt diseases (when grown with Red gram). Optimum temperature regime for chickpea is 24-30°C. Chickpea is a long day plant and requires sufficient bright sunshine. The low temperature duration during the growing season decides the duration of the crop, because of which in North India, it comes to harvest in 160-170 days. Whereas the winter is warm in South India then the duration is shorter of about 90-110 days. Among the micronutrients, zinc is an essential nutrient for the standard and healthy growth of plants. Generally, zinc affects the synthesis of protein in plants hence is the most critical micronutrient. Zn is also crucial in taking part in plant development due to its catalytic action in metabolism for all crops whereas Zn plays a vital role in growth of the plant in many of crucial processes such as synthesis of protein, structure and functions of membrane, expression of genes and oxidative stress tolerance. Similarly, application of $ZnSO_4$ significantly increased the chickpea yield. Therefore, the deficiency of Zn in soil leads to deficiency in crops and altogether this has become a problem all over the world with acute zinc deficiency ranges in the world. Among micronutrients, Zn deficiency is perhaps the most wide spread (Roy *et al.*, 2006) and (Ahlawat *et al.*, 2007) and is very common among chickpea growing regions of the world. Chickpea is generally considered sensitive to Zn deficiency, although there are differences in sensitivity to Zn deficiency between varieties. A comparison between several crop species has shown that chickpea is more sensitive to Zn deficiency than cereal and oil seeds. The critical Zn concentrations in soils vary from 0.48 mg kg^{-1} to 2.5 mg kg^{-1} depending on soil type and soils have low Zn availability when there is less than to 1.1 mg kg^{-1} of Zn (DTPA extraction). Zn deficiency decreases crop yield and delays crop maturity. Also, Zn deficiency reduces water use and water use efficiency and reduces nodulation, nitrogen fixation which contributes to a decrease in crop yield. In neutral to alkaline soils, Zn deficiencies can be encountered, Zn solubility decreases markedly above pH 6.0-6.5. Zn uptake is positively correlated with the amount of organic matter in the soil and Soils that have a higher concentration of sand and a low organic matter content produce lower crop yields which, leads to poor Zn utilization (Singh and Ram, 1996).

Shifting cultivation, or 'Jhum' cultivation as it is more commonly known in India, is any agricultural system in which fields are cleared by firing and are cropped discontinuously or an agricultural system which is characterized by a rotation of fields rather than crops by short periods of cropping (up to twenty and more years, but often as short as six to eight years), and by clearing by means of slash and burn. It is both a labour intensive and land extensive process of cultivation.

In North east about 2.7 million hectares are used for shifting cultivation. The number of years before a community returns to the same land depend on its availability to the community and the population, currently the cycle has shrunk to 5-7 years as compared to earlier decades. Generally, most of the hilly soils are found deficient of micronutrients like Zn, Fe and Mn. Therefore, the present research entitle "Comparative efficiency of zinc with and without organic manure on growth and effect of organic amendments in zinc nutrition of chickpea (*Cicer arietinum*) in Jhum soils of Manipur "was under taken with following objectives:

1. To determine zinc distribution in Jhum soils of Churachandpur district.
2. To determine the comparative efficiency of organic and inorganic sources in affecting the growth, yield and uptake of zinc by chickpea and
3. To estimate the critical limit of zinc in soils and crop.

Materials and Methods

Description of Study Site

This study was carried out in pots during *Rabi* season November 2016-17 at the research farm of college of agriculture, CAU, Imphal. Chickpea (variety GJG0891) was grown as test crop in pots with different treatment combination. The experimental site is located at Irroisemba (24° 51" N latitude, 93° 56" E longitude), in the NE - Hill Region of India. The area is 784 m above mean sea level and experienced an average of 1881 mm of precipitation annually with an average day temperature of 20° C.

Experimental Treatments and Design

a) Treatment Details

Plastic pots of 5kg capacity were used for the experiment. The pots were filled with 4kg of collected soils. 3 seeds of chick pea were sown in each pot, which were thinned out after 15 days of germination. The irrigation should be given thrice in a week. Plants were harvested at 90 days after sowing at the time of fully flowering stage, plant and soil samples were collected for analysis. The harvested plant samples were oven dried and grind then used for estimation of critical limits of zinc. The collected soil samples are used for the estimation of DTPA- extractable zinc

Soils from one sub-division (Churachandpur) will be collected for pot experiment with following treatments.

Table 5.1. Treatment Details

Treatment No.	*Zinc Sulphate with and without Vermicompost*
1	Zn @ 0 ppm from $ZnSO_4.7H_2O$
2	Zn @ 2.5 ppm from $ZnSO_4.7H_2O$
3	Zn @ 5 ppm from $ZnSO_4.7H_2O$
4	Zn @ 10 ppm from $ZnSO_4.7H_2O$
5	Zn @ 0 ppm mixed with vermicompost @ 5t/ha
6	Zn @ 2.5 ppm mixed with vermicompost @ 5t/ha
7	Zn @ 5 ppm mixed with vermicompost @5t/ha
8	Zn @ 10 ppm mixed with vermicompost @ 5t/ha
b) Layout and design	
Crop	*Chickpea*
Variety	GJG0809
Season	*Rabi* season
Experimental design	FRBD (Factorial randomized block design)
Number of replications	3
Number of treatments	8
Total number of pots	24×12=288 (12 villages from Churachandpur district)
Weight of soil per pot	4 kg

Soil Sampling and Plant Analysis

Soil samples of (0-15 cm depth) were collected from Churachandpur district covering 12 villages. From each selected villages five farmers were selected with the help of stratified at random sampling (proportional allocation) as given below:

Collection and Preparation of Soil Samples

District	*Sub-divisions*	*Number of Village*	*Respondent Farmers (each in one village)*	*Depth (cm)*	*Total*
Churachandpur	Churachandpur	12	5	0-15	60
Total number of samples from Churachandpur sub-division					60

Collected soil samples of the pots and field experiments brought into laboratory in a separate one kg polythene bag and dried at room temperature. After drying, broken clods were ground on wooden plank with wooden roller and passed through a 2 mm sieve. Thereafter, homogenized soil samples were stored in polythene bags for further analysis of selected physico-chemical properties of soil. For chemical analysis, soil samples were further ground and passed through a 0.5 mm sieve. Following methods were adopted for analysis of soil and plant samples:

Soil Property	*Methods of Analysis*
Mechanical Analysis	Jackson, 1973
Soil pH	Jackson, 1973
Electrical Conductivity	Jackson, 1973
Cation Exchange Capacity	Jackson, 1973
Organic Carbon	Jackson, 1973
Available Nitrogen	Jackson, 1973
Available Phosphorus	Jackson, 1973
Available Potassium	Jackson, 1973
DTPA – Extractable Zinc	Lindsay and Norvell, 1978

Zinc Content in Plant (mg kg^{-1})

The three plants sampled for estimating the zinc concentration from each pot at harvest were thoroughly washed with distilled water and dried in hot air oven at 65°C. Dried samples were powdered in a Willey mill to considerable fineness before storing them in polythene bags for further analysis. Powdered plant samples were treated with concentrated HNO_3 overnight for pre digestion. Then, the pre-digested samples were treated with di-acid mixture (HNO_3:$HClO_4$) (10:4 ratio) and digested on sand bath at low temperature till colourless white precipitate was obtained. The residue was dissolved in 6N HCl, filtered then the content was made to known volume by using 6N HCl. This was used for further nutrient analysis. The following analyses were carried out from the di-acid digested samples of whole plant and pod.

Micronutrient Zinc was estimated in the aliquot of plant extract using Atomic Absorption Spectrophotometer (Shimedzu model) as described by Jackson (1967).

Zn Uptake by Chickpea

Plant Zn uptake was obtained by multiplying of dry matter yield and plant Zn concentration.

$$\text{Nutrient uptake } (\mu g\ pot^{-1}) = \text{Nutrient content } (mg\ kg^{-1}) \times \text{Dry matter yield } g\ pot^{-1}$$

Critical Limit of Zinc in Soils and Plants

The critical limit of zinc in soil was determined by Bray's percent yield plotted against soil available zinc. Similarly, critical limits of zinc in Chickpea plant was determined by plotting Bray's percent yield of Chickpea against zinc concentration in plant, using the scattered diagram (Cate and Nelson 1965). Bray's percent yield was calculated using following formula:

Bray's per cent yield = Yield without zinc application ÷ Yield with optimum zinc level x 100

Statistical Analysis

Data obtained from the experiment will be subjected to statistical analysis with the help of analysis of variance technique for testing the significance of the treatment effects.

Brief Findings

Chickpea (*Cicer arietinum* L.) is one of the important crops of the world and is consumed by people across the globe. It is originated from Western Asia (Turkey). For vegetable purpose, it has become one of the most popular crop in the country. It can grow on wide range of soils from medium to heavy black soils. It does well on Black cotton soils and sandy loams. Optimum pH required for crop growth is 6.0 to 7.5 (>8.5pH not suitable). It does not withstand waterlogging, saline and alkaline conditions. Generally, zinc affects the synthesis of protein in plants hence is considered the most critical micronutrient.

Zn is also crucial in plant development due to its catalytic action in metabolism for all crops. Similarly, application of $ZnSO_4$ significantly increased the chickpea yield. Application of increasing levels of zinc with vermicompost significantly enhanced grain yield, straw yield, biological yield of legumes.

The mean distribution of zinc in surface soils of different villages of Churachandpur district recorded is 2.28 mg kg^{-1}, which is deficient compared to critical limits in soils as 2.60 mg kg^{-1}.

The results revealed that, the application of zinc with vermicompost significantly influenced plant height and number of branches from 40 DAS to upto harvest. The combined application of higher level of zinc with vermicompost T8 (Zinc 10 kg ha^{-1}+ 5 t ha^{-1}vermicompost) recorded higher growth parameters compared to lower level of zinc with and without vermicompost. The results revealed that, application of zinc with vermicompost significantly influenced dry matter production in chickpea. The combined application of higher levels of zinc with vermicompost T8 (Zinc 10 kg ha^{-1}+ 5 t ha^{-1}vermicompost) recorded higher total dry matter (19.60 g pot^{-1}) which was on par with combined applied zinc with and without vermicompost at lower levels. The Zn uptake was significantly increased with increase in its application from 2.5 to 10 kg $ha^{-1}$with and without vermicompost and a highest Zn uptake 289.17 μg $pot^{-1}$was recorded in treatment

T8 (Zinc 10 kg ha^{-1}+ 5 t ha^{-1}vermicompost). The critical limit of zinc concentration in soil was found to be 2.60 mg kg^{-1}. The critical limit of zinc concentration in chickpea plant at 90 days with the help of scattered diagram, critical level was found to be 12.20 mg kg^{-1}.

Conclusion

1. Based on physico chemical parameters of Jhum soils of Churachandpur district Manipur, are best suitable for chickpea cultivation with recommended doses of fertilizers along with zinc nutrition for better growth and yield of crop.
2. The comparative efficiency of zinc and zinc with vermicompost showing significantly increased on chickpea plant height (39.1 cm) and number of branches (3.00) per plant are recorded with increased level of zinc with vemicompost (Zinc 10 kg ha^{-1}+ 5 t ha^{-1}vermicompost).
3. And also, dry matter weight (11.87 g pot-1) and uptake of zinc (162.68-289.17 μg pot-1) are recorded with increased level of zinc with vemicompost (Zinc 10 kg ha^{-1}+ 5 t ha^{-1}vermicompost) in different villages of Jhum soils of Churachandpur.
4. For estimating the critical plant tissue concentration in chickpea, a value of 12.20 mg kg^{-1}is the critical limit. A value of 0.12 per cent plant tissue zinc concentration could distinguish the zinc deficient plants. Thus, the present study lays emphasis on zinc fertilization of chickpea plant on the basis of critical value in soil and plant.
5. It may be concluded that, the Jhum soils of Churachandpur having the soil physico chemical perameters (pH=6.64, EC=0.16 dS/m, OC=1.6 per cent, CEC=14.87 (C mol p+ kg^{-1}), N=414.23 (kg ha^{-1}), P_2O_5=14.25 (kg ha^{-1}), K_2O=141.08 (kg ha- 1), Zn=2.28 mg kg^{-1}) and critical limits of zinc in soil (2.60 mg kg^{-1}) and plant (12.20 mg kg^{-1}) in chickpea crop showing significant results in growth and dry matter yield with zinc 10 10 kg ha^{-1}alone and zinc with vermicompst (Zinc 10 kg ha^{-1}+ 5 t ha^{-1} vermicompost).

REFERENCES

1. Roy, R.N., Finck, A., Blair, G.J. and Tandon, H.L.S. (2006). Plant nutrition for food security. A guide for integrated nutrient management. FAO Fertilizer and Plant Nutrition Bulletin 16. Food and Agriculture Organization of the United Nations, Rome, Italy. 368 pp.
2. Ahlawat, I.P.S., Gangaiah, B. and Ashraf, Z.M. (2007). Nutrient management in chickpea. CAB Int., Wallingford,Oxon, United Kingdom, pp. 213 - 232.
3. Singh, M. and Ram, N. (1996). Effect of soil enrichment with zinc on crop yields and its replenishment in Mollisols of northern India. *Agro chemicals*, 40: 19-24.

4. Jackson, M.L. (1973). Soil chemical Analysis, Prentice Hall of India Pvt. Ltd,New Delhi. pp: 151-154
5. Lindsay, W.L. and Norvell, W.A. (1978). Development of ADTPA-Soil test for zinc, iron, manganese and cooper. *Soil Sci. Soc. of American J., 42*: 421-428.
6. Cate, R.B. and Nelson, L.A. (1965). Graphical procedure for critical limits of nutrients. *Proc. Soil Sci. Soc. Am.,* 89: 658.

Chapter 6

Comparative Efficacy of Different Manures on Chickpea (*Cicer arietinum* L.)

Vanlalmuanpuia Fanai

ABSTRACT

A pot experiment was conducted at College of Agriculture, Central Agricultural University, Imphal during Rabi season of 2017-18 to study the effect of organic manures (viz., FYM, compost, nitrogen enriched compost, phosphorus enriched compost and vermicompost) applied @ 5, 10 and 15 t ha^{-1}on the nutrient (nitrogen, phosphorus, potassium, zinc, iron, copper and manganese) release pattern, their uptake and yield of chickpea (JG-16) grown in a Typic Haploumbrepts soil. Results revealed that irrespective of different organic manures, significantly higher amount of available nutrients, plant nutrient concentration and their uptake were found in soil treated with manures over the untreated control at different stages of crop growth. Irrespective of different treatments available nutrients increased up to 20th day followed by a decline at harvest. The trend of nutrient uptake was gradual increase up to 80th day and then decreased at harvest. Nutrient uptake had a large influence on soil available nutrient accumulation. Comparing among the soils treated with manures significantly higher available nutrients was observed in soil treated with 15 t ha^{-1} followed by 10 t ha^{-1} and 5 t ha^{-1}at different sampling stages. Application of organic manures enhanced mineralization resulting rapid release of nutrients, thereby increasing plant nutrient accumulation, uptake and dry matter yield of chickpea. Significantly higher dry matter yield and yield (pot weight in g pot^{-1}) of chickpea was observed in soil treated with 15 t ha^{-1} followed by10 t ha^{-1} and 5 t ha^{-1}. Comparing among the organic manures applied at the same dose there was no significant *effect on nitrogen, potassium and micronutrient uptake, dry matter yield, yield (pod weight in g pot^{-1} and number of pods plant^{-1}) of chickpea. Comparatively higher yield (pod weight in g pot^{-1} and number of pods plant^{-1}) was observed in chickpea grown in vermicompost @ 15 t ha^{-1} (T17) applied soil which was at par with nitrogen enriched compost @ 15 t ha^{-1} (T15) and FYM @ 15 t ha^{-1} (T13), respectively. Application of organic manures is an effective way of maintaining available nutrient, their concentration, uptake by plant as well as crop yield. Application of vermicompost @ 15 t ha^{-1} will be more favourable for amending chickpea crop organically. Agronomic effectiveness of manures are more when applied at higher dose.*

Introduction

Chickpea (*Cicer arietinum* L.) is a winter pulse crop which comes under the sub-family Papilionaceae of the family Leguminosae. Indian gram is divided into two categories, *viz.*, Desi or brown gram (*Cicer arietinum* L.) and Kabuli or white gram (*Cicer kabulinum*). The former is most widely grown group. The stems are erect to spreading and have good branching ability. The latter have poor yield potential than desi type. In India, chickpea ranks first in area and production as well as consumption among pulse crops. Chickpea is mainly cultivated as winter crop in India, especially in the northern parts of the country. The area under chickpea cultivation is about 9.92 m ha and the yield in India is about 9.95 t ha^{-1} (FAO, 2014). Area under pulses in India, Pakistan, Nepal and Bangladesh covers 90 per cent of the world acreage (Saxena, 1987).

Production of chickpea in the whole world is as high as 13.37 million tonnes (FAO, 2014). The crop is grown for its pods. In India production of chickpea constitutes about 45 per cent of all the total pulses produced. Like other pulses, India remains the largest producer with about 8 million tonnes, accounting for about 67 per cent of total world production. Chickpea seeds contain about 17-20 per cent protein (Singh *et al.*, 2010). Though cultivated area of chickpea occupies an area to a great extent, the productivity is quite low comparing with its potential yield in the experimental station (Kumar, 1997). Imbalanced nutrition is one among many important factors which negatively affects the yield of chickpea. The supply of balanced nutrition can be done through inorganic fertilizers (chemicals) and organic manures. Inorganic fertilizers contain high amount of nutrients which remain readily available to plants. But the increasing cost of chemical fertilizers and depletion of soil health because of their application make the use of organics vital to enhance safe and efficient use in crop production. The soil organic carbon status is increased which makes it important for sustained crop yield. Soil biological, chemical and physical properties are adversely affected by continuous and indiscriminate use of chemical fertilizers, which in turn affects the sustainability of crop production (Virmani, 1994). Continuous use of high analysis fertilizers in an imbalanced manner leading to development of many problems like decline of soil organic matter, increasing salinity, sodicity, soil pollutant and hazards of pests and diseases (Chakraborti and Singh, 2004). Due to these reasons, many farmers prefer organic manures over inorganic fertilizers.

Nutrient management is also another important factor which plays a vital role affecting growth and yield of crops. In organic farming emphasis is given on maintenance and soil fertility recovery for sustainable yield of crops. Under organic production optimization of the nutrient requirement of crop increases its productivity. Soil productivity and fertility is enhanced through the management of organic matter under organic system. However, nutrients are slowly released from organic manures which may reduce crop yield and in turn leads to reduced net farm income which is a major drawback. Combined application of different organic manures produces higher yield and also improves soil health (Babalad, 1999).

Wastes from organic farm such as animal wastes, crop stubbles and litter can be utilized to prepare manures. The preparation of composts from livestock manures and farm wastes is an efficient way of recycling nutrients. In India, paddy straw is usually burned after harvest, or used as cattle feed. Straw burning creates pollution and more importantly causes nutrient loss. Instead of burning crop residues, they can be utilized for making manures such as compost, vermicompost, FYM, *etc.*, for crop nutrition (Mandal *et al.,* 1992; Dahiphale *et al.,* 2003; Singh and Nain, 2014).

Besides providing plant nutrients, organic manures improve the soil structure meanwhile enhancing beneficial soil microbe activities. Organic manures when incorporated in soil, through their effect on soil chemical properties, biological properties and physical properties enhance productivity (Graham *et al.,* 2000; Watson *et al.,* 2006; Pandey *et al.,* 2007). Organic manures help to develop favourable soil structure, minimize the soil temperature and maximize soil organic matter meanwhile releasing nutrients to the soil.

On an average, well decomposed FYM contains 0.5 per cent N, 0.2 per cent P_2O_5 and 0.5 per cent K_2O, compost contains 0.7 per cent N, 0.5 per cent P_2O_5 and 0.4 per cent K_2O, vermicompost contains 0.6-1.2 per cent N, 0.13-0.22 per cent P_2O_5 and 0.4-0.7 per cent K_2O (ISSS, 2012). The release pattern of nutrients differs with the type of supplied organic manures. The whole nutrient content may not be readily available to the crops when applied to soil. For enhancing crop productivity, it is important to know organic manure types which carry more available nutrients along with their release patterns.

Application of varying quantities of OM have significant effect on the contents of K, N and P of crops (Babu and Seshaiah, 2006; Vimala *et al.,* 2006; Azarmi *et al.,* 2008; Davari *et al.,* 2012; Bairwa and Yadav, 2017). Organic manures increase efficiency of nutrients by modifying their physical behaviour (Pandey *et al.,* 2007). Organic manures are locally available and can improve the soil characteristics and therefore they hold a great promise. In dryland areas and rainfed areas, organic manures can be used to replace organic fertilizers. (Badanur *et al.,* 1990). Crop amendment with organic manures enhance nutrient uptake. (Khankhane and Yadav, 2003; Rajkhowa *et al.,* 2003). Organic amendments increase nutrient concentration, nutritional value and nutrient balance of plant foods (Graham *et al.,* 2000; Beaulah *et al.,* 2002). The present agricultural production level is increased by efficient nutrient management. It sustains the production and it also protects the environment from hazards which may occur due to mismanagement of costly chemicals. The organic manure application to soil significantly enhanced yield parameters and the growth of crops (Patil *et al.*, 2012; Singh *et al.*, 2012).

Keeping the above facts in view, the present investigation entitled 'Comparative efficacy of different manures on chickpea (Cicer arietinum L.)' was taken up with the following objectives:

1. To study the nutrient release pattern of the different manures in soil
2. To study the effect of different organic manures on nutrient uptake and yield of chickpea.

Materials and Methods

Description of Study Site

A pot experiment entitled "*Comparative Efficacy of Different Manures on Chickpea (Cicer arietinum L.)*" was conducted in the Department of Soil Science and Agricultural Chemistry, College of Agriculture, Central Agricultural University, Imphal, during the *Rabi* season, 2017-18. All soil and plant analysis were done at the laboratory of Department of Soil Science and Agricultural Chemistry, CAU, Imphal.

Experimental Treatments and Design

a) Treatment Details

The experimental pots were prepared in the month of November, 2017. All the weeds, stones, stubbles and other materials were removed from the soil and made ready for plantation. 4 kg soils were filled in each 306 plastic pots. Recommended doses of fertilizer @ 20:40:20 kg NPK were applied as basal in the form of urea, DAP and MOP. The organic manures and fertilizers were applied to root zone of the soil in the pots five days prior to sowing. 5 chickpea (JG-16) seeds were sown in each pot and thinned out after 15 days of germination maintaining 3 plants per pot.

Rhizosphere soil samples were periodically collected on 0th, 20th, 40th, 60th, 80th days after sowing and at harvest for determination of available nutrients. The whole plants were also collected on 20th, 40th, 60th, 80th days after sowing and at harvest by destructive sampling to estimate nutrient uptake and concentration. The crop yields were also recorded. The experiment was carried out under RBD and consisted of 17 treatments with 3 replications each

b) Layout and Design

Soil Sampling and Data Analysis

Composite samples (0-20 cm depth) were collected from Central Agricultural University Research Farm, Andro, for the experiment. The soil samples were thoroughly dried in shade, ground and passed through 2mm sieve.

Five organic manures *viz.*, FYM, compost, nitrogen enriched compost, phosphorus enriched compost and vermicompost were prepared by taking paddy straw as raw material in the Department of Soil Science and Agricultural Chemistry, College of Agriculture, Central Agricultural University, Imphal. Cow dung slurry was prepared by mixing fresh cow dung and water at a ratio of 1:3 (1 kg of cow dung in 3 litres of water). Standard methods were adopted for preparation of FYM, compost, nitrogen enriched compost, phosphorus enriched compost and vermicompost.

Details of Treatments

Sl.No.	*Treatment*	*Amount Applied per Pot*
1	Control	Nil
2	Recommended dose of fertilizer (RDF) 0.017g urea,	0.154g DAP
	[20:40:20 NPK]	0.059g MOP
3	5 t/ha FYM	8.93g
4	5 t/ha Compost	8.93g
5	5 t/ha Nitrogen (urea) enriched compost	8.93g
6	5 t/ha Rock phosphate enriched compost	8.93g
7	5 t/ha Vermicompost	8.93g
8	10 t/ha FYM	17.86g
9	10 t/ha Compost	17.86g
10	10 t/ha Nitrogen (urea) enriched compost	17.86g
11	10 t/ha Rock phosphate enriched compost	17.86g
12	10 t/ha Vermicompost	17.86g
13	15 t/ha FYM	26.79g
14	15 t/ha Compost	26.79g
15	15 t/ha Nitrogen (urea) enriched compost	26.79g
16	15 t/ha Rock phosphate enriched compost	26.79g
17	15 t/ha Vermicompost	26.79g

Composition of Organic Manures

Organic Manures	*Total N (per cent)*	*Total P (per cent)*	*Total K (per cent)*
FYM	0.56	0.25	0.61
Compost	0.85	0.31	0.78
Nitrogen enriched compost	1.55	0.34	0.82
Phosphorus enriched compost	0.85	0.59	0.92
Vermicompost	1.22	0.50	1.38

Initial Properties of Soil before Growing the Crop

Parameters	*Results*
Textural	class Clay
Sand (per cent)	5.78
Silt (per cent)	22.56
Clay (per cent)	72.79
pH (1:2.5 soil: water ratio)	5.26
EC (1:2.5 soil: water ratio; dsm^{-1})	0.15
CEC [cmol (p+) kg^{-1}]	9.4
Organic carbon (per cent)	1.53
Available nitrogen (kg ha^{-1})	238.34
Available phosphorus (kg ha^{-1})	15.44
Available potassium (kg ha^{-1})	355.71

Following methods were adopted for analysis of soil and plant samples:

Soil Property	*Methods of Analysis*
Mechanical Analysis	Jackson, 1973
Soil pH	Jackson, 1973
Electrical Conductivity	Jackson, 1973
Cation Exchange Capacity	Jackson, 1973
Organic Carbon	Jackson, 1973
Available Nitrogen	Subbiah and Asija (1956).
Available Phosphorus	Bray and Kurtz (1945).
Available Potassium	Jackson, 1973
Available Micronutrients (Zn, Fe, Cu and Mn)	Lindsay and Norvell, 1978

Statistical Analysis

The data obtained from the experiment were analyzed by the analysis of variance technique for comparing the treatments as described by Gomez and Gomez (1984).

Brief Findings

Results of the experiment showed that in general the release pattern of macro (N, P, K) and micro (Zn, Fe, Cu, Mn) nutrient, their uptake by the crop and yield were affected by addition of different OM. Irrespective of different OM and doses of application the available nutrients increased up to 20th DAS which was then followed by a gradual decline in each successive sampling DAS *i.e.* 40th, 60th, 80th DAS and at harvest. Increase in the available nutrient might be accredited to enhanced mineralization which resulted in rapid release of nutrients. However, the declining trend with crop growth might be assigned to plant uptake. Significantly higher accumulation of available nutrients was found in soil treated with OM at different doses comparing with the untreated control. Further study showed that comparing among the soils treated with OM, available nutrients was found to be significantly highest in those soils which are treated with 15 t ha^{-1}which was then followed by 10 t $ha^{-1}$and 5 t ha^{-1}at different sampling stages.

At harvest, available N content was found to be significantly higher in soil treated with NEC @ 15 t ha^{-1}(T15) that was at par with VC @ 15 t ha^{-1}(T17). Soil added with PEC @ 15 t ha^{-1}(T16) showed higher available P but was at par with VC @ 15 t ha^{-1}(T17) and available K was statistically higher in VC @ 15 t ha^{-1}(T17) treated soil that was at par with NEC @ 15 t ha^{-1}(T15) at harvest. Available Zn content in soil after the harvest of crop was shown highest in soils treated with FYM @ 15 t ha^{-1}(T13) that was at par with VC @ 15 t ha^{-1}(T17). Concentration of available Fe and Mn in soil after harvest of crop was found to be significantly higher in soils treated with VC @ 15 t ha^{-1}(T17). At the time of harvest, available Cu in soil was higher in soils which were treated with PEC @ 15 t ha^{-1}(T16) that was at par with VC @ 15 t ha^{-1}(T17).

Results further showed that irrespective of different treatments given to the crops, amount of total nutrient concentration in plant increased up to 40th DAS seed and then gradually declined up to harvest. The application of OM significantly enhanced plant N, P and K concentration in comparison with control at the time of harvest. Comparing with the untreated control, OM application to soil was found to have significant effect on total Zn, Fe, Cu and Mn content of the crop at different growth stages. This showed that application of OM promoted plant nutrient accumulation. The application of OM at different doses did not have significant effect on the total nutrient content of chickpea at different crop growth stages.

After harvest, comparing among the OM amended soils, highest N concentration was recorded in soil which were treated with NEC @ 15 t ha^{-1}(T15) that was at par with VC @ 10 t ha^{-1}(T12). Significantly higher P concentration was seen in soils treated with PEC @ 15 t ha^{-1}(T16) that was at par with PEC @ 10 t ha^{-1}(T11). Soils treated with VC @ 10 t ha^{-1}(T12) accumulated higher K concentration in chickpea but stayed on par with NEC @ 15 t ha^{-1}(T15) at harvest. It was also seen that at harvest significantly higher Zn and Mn concentrations were recorded in those crops grown in FYM @ 10 t ha^{-1}(T8) that was at par with NEC @ 10 t ha^{-1}(T12) and compost @ 15 t ha^{-1}(T14), respectively. Chickpea crops grown in those soil treated with NEC @ 10 t ha^{-1}(T12) accumulated higher Fe but stayed on par with PEC @ 5 t ha^{-1}(T6). At harvest, comparing between the soils amended with OM, the highest Cu content was recorded in NEC @ 15 t ha^{-1}(T15) and VC @ 15 t ha^{-1}(T17).

Result showed that irrespective of different treatments, the trend of macronutrient uptake and micronutrient uptake by the crop showed an increase up to 80th DAS which was then followed by a decline at harvest. With increase in the production of dry matter, uptake of nutrient by crops increased and this may be accredited to increased nodule count and high rate of mineralization of OM added to the soil. In general, application of OM at different doses showed significant effect on uptake of nutrient by the crop in comparison with control at different crop growth stages. More nutrient uptake could be credited to improvement in the physical properties, chemical properties, biological properties and microbiological properties of soil when OM are added as they are effective source of nutrients.

In general, comparing between the OM applied at the same dose, no significant effect on N, K and micronutrient uptake was shown at different growth stages of chickpea. Study showed that at later crop growth stages *i.e.* 60th, 80th DAS and at harvest application of OM @ 10 t ha^{-1}showed significantly higher Zn uptake than the same OM applied at 5 t ha^{-1}. Comparing between the OM amended soils, the highest N, P and K uptake was seen in soil added with NEC @ 15 t ha^{-1}(T15) that was at par with PEC @ 15 t ha^{-1}(T16), VC @ 15 t ha^{-1}(T17) and VC @ 10 t ha^{-1}(T12), VC @ 15 t ha^{-1}(T17) respectively. At harvest, highest Zn uptake was found in soil treated with VC @ 15 t ha^{-1}(T17) that was at par with that of compost @ 15 t ha^{-1}(T14). Fe and Mn uptake was higher in NEC @ 15 t ha^{-1}(T15) treated soil that was statistically at par with VC @ 10 t ha^{-1}(T12) and VC @ 15 t ha^{-1}(T17). Comparing among the OM, Mn uptake was found to be significantly higher in soil added with compost @ 10 t ha^{-1}(T14) but stayed on par with FYM @ 10 t ha^{-1}(T8) at the time of harvest.

Application of different doses of OM increased yield (dry matter) of chickpea up to 80th DAS and then decreased at harvest. Dry matter yield of the crop grown in OM amended soils was found significantly higher than control on 60th, 80th DAS and at harvest. Highest yield in dry matter of chickpea was seen in soils treated with 15 t ha^{-1} which was then followed by 10 t ha^{-1} and 5 t ha^{-1}. The increase in yield of dry matter with increasing levels could be assigned to the increase in available nutrients which are released from the OM. Comparing among the different sources NEC @ 15 t ha^{-1} (T15) resulted in higher yield of dry matter that was at par with the VC applied at the same dose (T17) at harvest.

Comparing between OM amended soils, pod yield (g pot^{-1}) was found to be significantly higher in soils treated with 15 t ha^{-1} which was then followed by 10 t ha^{-1} and 5 t ha^{-1}. However, different doses of treatments did not show significant differences in the pod yield (no. of pods plant^{-1}) of the crop. No significant difference was shown in the yield (pod weight in g pot^{-1} and no. of pods plant^{-1}) between the soils which were treated with OM at the same dose. Significantly higher yield (pod weight in g pot^{-1} and no. of pods plant^{-1}) was seen in chickpea grown in VC @ 15 t ha^{-1} (T17) added soil that was at par with NEC @ 15 t ha^{-1} (T15) and FYM @ 15 t ha^{-1} (T13) respectively.

Conclusion

From the results collected from the investigation, it could be concluded that application of OM is an effective way of maintaining the quantity of available macro and micronutrient in soil. Addition of OM enhanced nutrient concentration and uptake by plant. The application of OM improved plant growth along with the yield attributes of the crop. Nutrient availability and agronomic effectiveness of OM are more when they were applied at higher dose. It can be concluded that application of VC @ 15 t ha^{-1}was found to be more favorable for amending chickpea crop organically.

REFERENCES

1. Food and Agriculture Organization of the United Nations (FAO) (2014). Agricultural production year book. Regional office for Asia and the pacific, Bangkok.
2. Saxena, M.C. (1987). Problems and potential of chickpea production in the nineties. Proceeding of the 2nd Int. Workshop on Chickpea Improv., pp. 13-25.
3. Singh, N.P., Sewak, S., Iquebal, M.A. and Nath, O. (2010). Research Highlights: All India Coordinated Research Project on Chickpea. IIPR, Kanpur.
4. Kumar, J. (1997). Missed opportunities. In: The Hindu Survey of Indian Agriculture, pp. 49-53.
5. Virmani, S.M. (1994). The twenty first - Dr. R.V. Tamhane memorial Lecture: UNCEED Agenda 21: The new challenge for soil research. *J. Indian Soc. Soil Sci.*, 42: 516-523.
6. Chakarborti, M. and Singh, N.P. (2004). Bio-compost: a novel input to organic farming. *Agrobios Newsletter*, 2(8): 14-15.
7. Babalad, H.B. (1999). Integrated nutrient management for sustainable production in soybean based cropping systems. Ph.D Thesis, Submitted to University of Agricultural Sciences, Dharwad.
8. Mandal, R.K., Subuddhi, U.K. and Singh, Y.V. (1992). Response of rice (*Oryza sativa*) to green-leaf manure and prilled urea. *Indian J. Agril. Sci.*, 6(7): 432-450.
9. Dahiphale, A.V., Giri, D.G., Thakre, G.V. and Gin, M.D. (2003). Effect of integrated nutrient management on yield and yield contributing parameters of the scented rice. *Annal. Pl. Physiol.*, 17: 24-26.
10. Singh, S. and Nain, L. (2014). Microorganisms in the conversion of agricultural wastes to compost. *Proc. Indian Natn. Sci. Acad.*, 80(2): 473-481.
11. Graham, R.D., Humphrius, J.M. and Kitchen, J.L. (2000). More enhanced cereals: a sustainable foundation for diet. *Asia Pacific J. Clin. Nutr.*, 9: 9.
12. Watson, C.A., Atkinson, D., Gosling, P., Jackson, L.R. and Rayns, F.W. (2006). Managing soil fertility in organic farming systems. Soil Use and Management, 18: 239-247.
13. Pandey, N., Verma, A.K., Anurag and Tripathi, R.S. (2007). Integrated nutrient management in transplanted hybrid rice (*Oryza sativa* L.). *Indian J. Agron.*, 52(1): 40-42

14. Babu, G.K. and Seshaiah, B.V. (2006). Effects of phosphates on yield and nutrient uptake in paddy. *Agropedology*, 16(1): 50-53.

15. Vimala, P., Illias, M.K. and Salbiah, H. (2006). Effect of rates of organic fertilizer on growth, yield and nutrient content of cabbage (*Brassica oleraceae* var. capitata) grown under shelter. *Acta-Horticulturae*, 7(10): 391-397.

16. Azarmi, R., Giglou, M.T. and Taleshmikail, R.D. (2008). Influence of vermicompost on soil chemical and physical properties in tomato (*Lycopersicum esculentum*) field. *African J. Biotechnol.*, 7(14): 2397-2401.

17. Davari, M.R., Sharma, S.N. and Mirzakhani, M. (2012). The effect of combinations of organic materials and biofertilizers on productivity, grain quality, nutrient uptake and economics in organic farming of wheat. *J. Org. Sys.*, 7(2): 26-35.

18. Bairwa, S. and Yadav, P.K. (2017). Influence of FYM, inorganic fertilizers and micronutrients on soil nutrient status and plant nutrient contents and their uptake by African marigold (*Tagetes erecta* Linn.). *Int. J. Curr. Microbiol. App. Sci., 6*(6): 1362-1370.

19. Badanur, V.P., Poleshi, C.M. and Naik, B.K. (1990). Effect of organic matter on crop yield, physical and chemical properties of a Vertisol. *J. Indian Soc. Soil Sci.*, 38: 429-430.

20. Khankhane, P.J. and Yadav, B.R. (2003). Comparative manurial performance of FYM, biogas slurry and sewage sludge. *Ann. Agric. Res. New Series*, 24(1): 148-150.

21. Rajkhowa, D.J., Saikia, M. and Rajkhowa, K.M. (2003). Effect of vermicompost and levels of fertilizer on green gram. *Legume Res.*, 26(1): 63-65.

22. Graham, R.D., Humphrius, J.M. and Kitchen, J.L. (2000). More enhanced cereals: a sustainable foundation for diet. *Asia Pacific J. Clin. Nutr.*, 9: 9

23. Beaulah, A., Vadivel, E. and Rajadirai, K.R. (2002). Studies on the effect of organic manures and inorganic fertilizers on the quality parameters of Moringa (*Moringa oleifera* Lam) cv. PKM1. In: Abstracts of the UGC Sponsored, Nat. Sem. On Emerging Trends in Horticulture, Department of Horticulture, Annamalai Univ., Annamalai Nagar, Tamil Nadu, pp. 127-128.

24. Patil, S.V., Halikatti, S.I., Hiremath, S.M., Babalad, H.B., Sreenivasa, M.N., Hebsur, N.S. and Somanagouda, G. (2012). Effect of organics on growth and yield of chickpea (*Cicer arietinum* L.) in Vertisols. *Karnataka J. Agric. Sci.*, 25(3): 326-331.

25. Singh, G., Sekhon, H.S. and Kaur, H. (2012). Effect of vermicompost, farm yard manure and chemical nutrients on growth and yield of chickpea (*Cicer arietinum* L.). *Int. J. Agric. Res.*, 7(2): 93-99.

26. Jackson, M.L. (1973). Soil Chemical Analysis. Prentice Hall of India Pvt. Ltd., New Delhi.

27. Subbiah, B.V. and Asija, G.L. (1956). A rapid procedure for estimation of available N in soils. *Curr. Sci.,* 25: 259-260.
28. Bray, R.H. and Kurtz, L.T. (1945). Determination of total, organic and available forms of phosphorus in soils. *Soil Sci.*, 59: 39-45.
29. Lindsay, W.L. and Norvell, W.A. (1978). Development of DTPA soil test for zinc, manganese and copper. *Soil Sci. Soc. Am. J.*, 42: 421-428.
30. Gomez, K.A. and Gomez, A.A. (1984). Statistical procedures for agricultural research. John Wiley and sons. Inc. London, UK (2nd edition)

Chapter 7

Nitrogen Transformation in Soil Amended with Humic Acid Derived from Organic Manures

Geeta Nongmeikapam

ABSTRACT

An experiment was conducted in the Department of Soil Science and Agricultural Chemistry, College of Agriculture, Central Agricultural University, Imphal during Rabi season of 2017-18 to study the effect of humic acid (HA) derived from organic manures (viz., FYM, compost, Vermicompost and N- enriched compost) on changes in different inorganic and organic forms of nitrogen in soil and N uptake of chickpea (var. JG-16) with the treatments: T1 - control,T2 - Recommended dose of nitrogen (RDN) through urea,T3 - RDN through FYM,T4 - RDN through compost, T5 - RDN through VC, T6 - RDN through N enriched compost, T7 - RDN through HA (FYM), T8 - RDN through HA (Compost), T9 - RDN through HA (VC) and T10 - RDN through HA (N enriched compost). Results revealed that irrespective of different nitrogen sources, significantly higher amount of inorganic and organic forms of nitrogen was found in soil treated with nitrogen over the untreated control. Comparatively higher amount of exchangeable NH_4^+, soluble NO_3^-, unidentified hydrolysable organic N, N concentration, dry matter yield, N uptake and pod yield (pod weight in g pot^{-1} and no. of pods pot-1) were observed in soil treated with organic manures as compared to the corresponding humic acids. However, higher amount of hydrolysable organic N fractions (hydrolysable NH_4^+, hexosamine, amino acid and serine + threonine - N), total hydrolysable organic N, non - hydrolysable organic N and total N were accumulated in humic acid amended soil as compared to the corresponding organic manures. Irrespective of different treatments, total N and total hydrolysable organic N *gradually declined till harvest comparing with initial value. Among the different hydrolysable organic N, hydrolysable NH_4^+ - N, hexosamine - N and amino acid - N get mineralized and give main contribution to available N. Significant contribution of hydrolysable forms of N towards available N in the soil is a positive indication for substitution of chemical N by organic sources. Organic manures mineralized at faster rate with higher efficiency comparing with the corresponding humic acid. Among the different organic manures, faster mineralization takes place in FYM. Comparing between the different organic N sources, significantly higher N uptake by chickpea and yield (no. of pods pot^{-1}and pod weight in g pot-1) was found in FYM treated soil.*

Introduction

Nitrogen (N) is one of the most important plant nutrient required by plants for growth and metabolic activities. In soil, nitrogen exists in two major forms: organic and inorganic. A bulk of total nitrogen is present in the organic form and only about 2 per cent in inorganic form (Rattan, 2009). The organic form of nitrogen, particularly the hydrolysable form, is slowly mineralized and is transformed to mineral nitrogen by soil microbes (Chakravorty *et al.,* 1992; Ling - Ling and Shu - Tian, 2014). During the transformation, nitrogen is converted from one form to another (Broadbent, 1965; Chakravorty *et al.,* 1992; Ling - Ling and Shu - Tian, 2014). Bremner (1965) reported that more than 95 per cent of the total N was in organic combination, a portion of which, on hydrolysis, is liberated as ammonium (10-25 per cent), a greater portion as amino acids (25-40 per cent) and a smaller fraction as amino sugar (1-5 per cent). These hydrolysable N fractions are the main source N availability for the nutrition of organisms in soil.

Nitrogen, present in soil, mostly comes from natural sources and applied fertilizers. Natural sources, whether it is from soil organic matter or added organic matter, are of utmost importance with respect to maintenance of soil fertility as a whole and controlled release of nitrogen during the cropping season. It is critical to understand what happens to organic nitrogen after soil application in order to use it efficiently for crop production. N-transformation in soil is influenced by organic matter additions (Sihag *et al.,* 2005; Abbasi *et al.,* 2007; Qian- Ru *et al.,* 2009; Tabassum *et al.,* 2010; Kaur and Singh, 2014). Its release from organic matter depends upon its source and state of decomposability (Khalil *et al.,* 2001; Sajjad *et al.,* 2003; Nourbakhsh and Dick, 2005; Mubarak *et al.,* 2010; Baitilwake *et al.,* 2012). Helgason *et al.* (2007) reported that less than 5 per cent of organic N content of composted manures was mineralized over 425 days, suggesting that little of the organic N becomes available in the year of application. However, faster mineralization of manures was also reported with an initial rapid N release at 0-30 days, phase of constant release at 45- 55 days, decline phase at 70-90 days and sharp increase in N release at 120 days by Azeez and Averbeke (2010) and Baitilwake *et al.* (2012). Crops depend on inorganic forms of nitrogen which are released after the mineralization of organic nitrogen in soil. Unfortunately, organic nitrogen in soil is not as amenable to manage as is fertilizer from the bag. The organic N cannot be locally concentrated, applied at just the optimum time or measured out in just the right amounts. If the way the mineralization process responds to variables in soil and climate is understood, it may be possible to adjust variety, planting time and moisture addition to take advantage of the peculiarities of a given soil and environment.

Nitrogen is an essential component of all amino acids which are present in humic acid (HA). Humic acid is the major constituent of humic substances and it is an integral part of organic matter and one of the key component of terrestrial ecosystem (Paul and Saha, 2016). Humic acids are primarily composed of carbon (52 per cent), oxygen (32 per cent), nitrogen (1-5 per cent) and hydrogen (8 per cent). Furthermore, humic acid fraction of humic substances contain 20 to 50 per cent amino acid nitrogen, 3 to 4 per cent amino sugar nitrogen, small amount of purine and pyrimidine and rest unidentified (Flaig *et al.,* 1975). Its content of the humic acid varies from 1-5 per cent (Flaig, 1975) depending upon its origin and sources of organic matter. Nitrogen present in humus can serve as a slow release nitrogen fertilizer (Flaig and Sochtig, 1967 and Schnitzer and Khan, 1972; Filip and Kubat, 2003 and Nguyen *et al.,* 2004a). However, reports on more susceptible nature of humic acid to mineralization than organic manures were given by Satisha and Devarajan (2011) and Devi and Saha (2017).

Natural organic substances such as humic and fulvic acids play an essential role in ensuring soil fertility and plant nutrition. Addition of such molecules either to the soil or through foliar spray along with adequate amount of conventional fertilizers improves the efficiency of applied fertilizers apart from promoting the conversion of unavailable form of nutrients to available forms. The application of humic acids was found to increase the content and total amount of nitrogen in plant (Govindasamy and Chandrasekaran, 1992; Sharif *et al.,* 2002). Humic acids in small amounts act as specific sensitizing agents increasing the permeability of the cell membrane, which results in increased uptake of nutrients by the plants in large amounts (Senn and Kingman, 1973). Dry matter production and N uptake by plant grown in soil applied with organic manures showed higher than the chemical fertilizer (Helgason *et al.,* 2007). Irrespective of sources, N play akey role in agriculture by increasing crop yield (Ahmad *et al.,* 2008). Immobilization of N due to application of organic manures and humic acid was reported by Bird *et al.* (2002) and Mohanty *et al.* (2011). So, understanding the effect of organic amendments on the transformation of nitrogen into different forms is a pre-requisite for managing nitrogen input for a given soil.

Keeping the above points in view the present investigation entitled "*Nitrogen Transformation in Soil Amended with Humic Acid Derived from Organic Manures*" has been undertaken with the objectives:

1. To study the changes in different inorganic and organic forms of nitrogen in soil amended with humic acid (HA) derived from organic manures.

2. To study the effect of humic acid derived from different organic manures on nitrogen uptake of chickpea.

Materials and Methods

Description of Study Site

Composite soil samples (0-15 cm depth) were collected from Central Agricultural University Research farm, Andro for the experiment. The soil belongs to *Typic Haploumbrepts* (Sen *et al.,* 2006). The collected soil samples were air dried at room temperature in shade. After drying, the clods were ground with wooden roller on a wooden plank and passed through a 2mm sieve. The sieved homogenized soil samples were stored in labeled polythene bags for the experiment.

Preparation of Organic Manures

Four organic manures (*viz.,* FYM, compost, vermicompost and N - enriched compost) were prepared taking paddy straw as raw material in the Department of Soil Science and Agricultural Chemistry, College of Agriculture, Central Agricultural University, Iroisemba, Imphal.

Preparation of Compost

Four kg of chopped (4-5 cm) paddy straw sprinkled with water were filled in a thermocol box and kept for pre-decomposition. After pre-decomposition, alternate layers of paddy straw and cow dung slurry (1kg of cow dung in 3 litres of water) were arranged. Holes were made at the four sides and bottom of the box for proper aeration. Turning was done at an interval of 15 days for maintaining aeration and proper decomposition. Compost get matured after 110 days.

Preparation of FYM

Four kg of chopped (4-5 cm) paddy straw was spread on the floor of the cattle shed so as to soak with the urine and excreta of the cattle. The soaked materials along with cow dung were collected in a thermocol box and plastered with thin layer of mud. FYM get ready to use within about 135 days.

Preparation of Vermicompost

Vermicomposting was carried out after pre- decomposition of 4kg wetted paddy straw (4-5 cm) for one week. Side holes were created on the thermocol box to enhance aeration. Four holes were also made at the bottom of the thermocol box and covered with muslin cloth. The pre- decomposed paddy straw (15 cm) was spread evenly above the cowdung slurry (1kg of cow dung in 3 litres of water) in sandwich manner. The top layer of paddy straw was covered with cow dung slurry and twelve matured vermiworms (*Eisenia foetida*) were released. The box was then covered with wetted gunny bags to maintain moisture and enhance decomposition. Turning was done at an interval of 15 days. Watering was stopped before a week of full maturity of the vermicompost and it was fully matured after 110 days. The matured vermicompost was made into dome shape and harvested. It was then air dried in shade and sieved for use.

Preparation of N-enriched Compost

Four kg of partially decomposed chopped (4-5 cm) paddy straw were placed in a thermocol box and arranged in four layer with cow dung slurry alternately. Urea was added @ 20g kg^{-1}raw material and kept moist by covering with gunny bags. Turning was done at 15 days interval for maintaining aeration and proper decomposition. Compost get ready after 110 days. The prepared FYM, compost, vermicompost and N - enriched compost were air dried, ground, passed through 2mm sieve, kept in airtight plastic bags and used for the experiment.

Extraction of Humic Acid

Humic acids were extracted from FYM, compost, vermicompost and N - enriched compost following the procedure of Stevenson (1996). Twenty gram (20 g) of air dried, ground and passed through 2mm sieve organic manure were placed in 250 ml conical flask and 200 ml of 0.5 N NaOH solution were added, shaken for 12 hours on a mechanical shaker and transferred to a buckner funnel using whatman No 42 filter paper. Concentrated HCl were added drop wise and stirred with a glass rod until cloudiness appeared in the solution and let it sit overnight to complete the precipitation of humic acid. The solution were then centrifuged at 5000 rpm for 20 min. The humic acid were collected on the filter paper and dialyzed overnight to make ion salt free. The humic acid were then dried in oven, ground, passed through 2mm sieve, kept in airtight container and used for the investigation.

Experimental Setup

Four kg soils were filled in each one hundred eighty (180) plastic pots. Each pot was sown with 5 seeds of chickpea (JG-16) and thinned out after 15 days of germination maintaining one plant per pot. Recommended doses of phosphorus and potassium at 40 and 20 kg ha^{-1}were applied as basal in the form of DAP and MOP. Recommended dose of nitrogen (20 kg ha^{-1}) was added through urea, different organic manures (FYM, compost, vermicompost, N - enriched compost) and humic acids derived from the corresponding organic manures maintaining different sets of treatments. Urea, organic manures and humic acids were mixed thoroughly with the soil. Then soils of each treatment were moistened to 60 per cent of water holding capacity (WHC). The loss of moisture were replenished by the periodic addition of sterile distilled water on every alternate day by difference in weight. Total N content of the organic N sources and equivalent weight based on the recommended dose of nitrogen taken for different sources are shown in Table 7.

Organic manures from different sources *viz.,* farmyard manure (FYM), compost, vermicompost, and N- enriched compost and humic acid extracted from the organic manures were prepared, air dried, ground and passed through 2mm sieve.

Table 7.1: Total N Content and Equivalent Weight (g pot^{-1}) of the N Sources

N Sources	*Total N Content*	*Equivalent Weight Based on*
(Per cent)	*Recommended Dose of N (g pot^{-1})*	
i. FYM	0.52	6.86
ii. Compost	0.80	4.46
iii. Vermicompost	1.08	3.31
iv. N- enriched compost	1.49	3.29
v. HA from FYM	2.14	1.67
Vi. HA from compost	0.94	3.81
vii. HA from vermicompost	1.84	1.95
viii. HA from N – enriched Compost	2.8	1.28

Rhizosphere soil samples were periodically collected on 0th, 20th, 40th, 60th, 80th days after sowing seeds and at harvest by for determination of exchangeable NH_4^+, soluble NO_3^-, total N, total hydrolysable organic N, hydrolysable organic NH_4^+, hexosamine - N, amino acid - N and (serine + threonine) - N. The whole plants were collected on 0th, 20th, 40th, 60th, 80th days after sowing seeds and at harvest by destructive sampling to estimate N uptake. Crop yield were also recorded. The experiment was carried out under CRD with 10 treatments replicated thrice.

Treatment Details

T1 = (control)

T2 = RDN through urea

T3 = RDN through FYM

T4 = RDN through compost

T5 = RDN through VC

T6 = RDN through N enriched compost

T7 = RDN through HA (FYM)

T8 = RDN through HA (Compost)

T9 = RDN through HA (VC)

T10 = RDN through HA (N enriched compost)

- ☆ *VC*- Vermicompost
- ☆ *FYM*- Farm yard manure
- ☆ *RDN* - Recommended Dose of Nitrogen
- ☆ *HA* - Humic Acid

Table 7.2. General Characteristics of the Soil Used in the Experiment

Soil Characteristics	*Results*
i. Textural class	Clay
ii. Sand (per cent)	5.78
iii. Silt (per cent)	22.56
iv. Clay (per cent)	72.79
v. Maximum water holding capacity (per cent)	44.71
vi. pH (1:2.5 soil: water ratio)	5.26
vii. EC (1:2.5 soil: water ratio, dsm-1)	0.15
viii. CEC [cmol(p+)kg^{-1}]	9.40
ix. Organic carbon (per cent)	1.53
x. Total nitrogen (mg kg^{-1})	1191.13
xi. Available nitrogen (mg kg^{-1})	84.00
xii. Available phosphorus (mg kg^{-1})	16.79
xiii. Available potassium (mg kg^{-1})	17.50

Following methods were adopted for analysis of soil and plant samples:

Soil Property	*Methods of Analysis*
Mechanical Analysis	Jackson, 1973
Maximum Water Holding Capacity	Saha, 2008
Soil pH	Jackson, 1973
Electrical Conductivity	Jackson, 1973
Cation Exchange Capacity	Jackson, 1973
Organic Carbon	Jackson, 1973
Available Nitrogen	Subbiah and Asija (1956).
Available Phosphorus	Jackson, 1973
Available Potassium	Jackson, 1973
Total Nitrogen	Jackson, 1973
Exchangeable NH_4^+ and Soluble NO_3^-	Bremmer and Keeney, 1966
Organic Nitrogen Fractions (Includes Total Hydrolysable – N, Ammonium -N, Ammonium + Hexosamine – N, Hexosamine – N, Serine + Threonine – N, Amino Acid – N, Unidentified – N, Non-hydrolysable – N)	Stevenson, 1966

Statistical Analysis

Data of all the inorganic and organic forms of nitrogen at each stages were statistically analysed through analysis of variance technique for comparing the treatments effects as described by Gomez and Gomez (1984).

Brief Findings

N is a major element required for proper growth and development of crop. It is present in soil in both organic and inorganic form. Soil application of N can be done either through inorganic or organic sources. The amount of N present in organic matters actually controls the release pattern of N in soils and soil mineral N dynamics. N-transformation in soil differs due to addition of either organic matter or humic acid. Addition of organic matter of different decomposition rate as a source of N is well established but direct application of humic acid as a source of N to a soil has not been much studied in detail. Therefore, it is of utmost importance to study N transformation in soil treated with humic acids derived from organic manures of different sources.

Considering the above points, a pot experiment entitled '*Nitrogen Tranformation in Soil Amended with Humic Acid Derived from Organic Manures*' was carried out in the Department of Soil Science and Agricultural Chemistry, College of Agriculture, Central Agricultural University, Iroisemba, Imphal during the Rabi season, 2017-18) to study the effect of humic acid (HA) derived from organic manures (*viz.,* FYM, compost, vermicompost and N - enriched compost) on changes in different inorganic and organic forms of nitrogen in soil and N uptake of chickpea (var. JG-16) with the treatments: T1 - control,T2 - Recommended dose of nitrogen (RDN) through urea,T3 - RDN through FYM,T4 - RDN through compost, T5 - RDN through VC, T6 - RDN through N enriched compost, T7 - RDN through HA (FYM), T8 - RDN through HA (Compost), T9 - RDN through HA (VC) and T10 - RDN through HA (N enriched compost). The salient findings of the investigation are summarized and concluded here under:

Results showed that in general, changes in the accumulation of different inorganic as well as organic forms of nitrogen were affected greatly by organic matter addition. Irrespective of different nitrogen sources, significantly higher amount of inorganic and organic forms of nitrogen was found in soil treated with nitrogen over the untreated control. Result further revealed that the amount of exchangeable NH_4^+ -N decrease up to 20th day for soil treated with urea and organic manures and then gradually increase till 60th day with a decline at harvest. However, a different trend was observed for soil treated with humic acids derived from organic manures showing a decline upto 40th day followed by an increase till 80th day and again decreased at harvest. Irrespective of different treatments, the amount of soluble NO_3^- gradually increase up to 20th day with a slight decrease at 80th and finally increase at harvest in humic acid treated soils. However, soil treated with organic manures presented different trend with an increase at 20th day followed by a decline at 40th, again increase up to 60th day and then decrease at 80th with an increase at harvest. Irrespective of different treatments, total N and total hydrolysable organic N gradually declined till harvest comparing with initial value. This shows more vulnerable nature of total hydrolysable organic N to mineralization and could be considered as a major source of potentially available N for plants. The amount of hydrolysable organic NH_4^+ and amino acid decreased gradually up to 40th with an increase up to 80th and finally declined at harvest. The amount of non- hydrolysable organic - N and unidentified hydrolysable organic N gradually increase up to 40th day followed by a decrease till harvest for all N treatments. The trend of changes in N concentration in chickpea grown in organic manures and humic acid treated soils occurred similarly with gradual increase up to 40th followed by a decrease up to 80th and finally increased at harvest. There was an increasing trend of dry matter yield of chickpea up to 80th days with slight decline at harvest. N uptake by chickpea grown in soil amended with organic manures and humic acid showed gradual increase up to 80th day followed by a decrease till harvest.

Comparatively higher amount of exchangeable NH_4^+, soluble NO_3^-, unidentified hydrolysable organic N, N concentration, dry matter yield, N uptake and pod yield (pod weight in g $pot^{-1}$and no. of pods pot^{-1}) were observed in soil treated with organic manures as compared to the corresponding humic acids. However, higher amount of hydrolysable organic N fractions (hydrolysable NH_4^+, hexosamine, amino acid and serine+threonine -N), total hydrolysable organic N, non - hydrolysable organic N and total N were accumulated in humic acid amended soil over the corresponding organic manures.

Conclusion

Based on the results obtained from the investigation carried out, it could be concluded that application of organic manures is an effective way of maintaining the quantity and quality of soil inorganic and organic N. Among the different hydrolysable organic N, hydrolysable NH_4^+ - N, hexosamine - N and amino acid - N get mineralized and give main contribution to available N. Significant contribution of hydrolysable forms of N towards available N in the soil is a positive indication for substitution of chemical N by organic sources. Organic manures mineralized at faster rate comparing with the corresponding humic acid. Among the different organic manures, faster mineralization takes place in FYM with higher efficiency. Comparing between the different organic N sources, significantly higher N uptake by chickpea and yield (no. of pods pot^{-1} and pod weight in g pot^{-1}) was found in FYM treated soil.

REFERENCES

1. Rattan, R.K. (2009). *Fundamentals of Soil Science,* Indian Society of Soil Science, National Agricultural Science Centre Complex, Dev Prakash Sastri Marg, Pusa, New Delhi.

2. Chakravorty, P.K and Mandal, L.N. (1992). Effects of organic and chemical sources of nitrogen on nitrogen transformation, growth and yield of rice under submerged condition. *Ind. Agric.,* 36,131 -140.

3. Ling-Ling, L.I. and Shu-Tian, L.I. (2014). Nitrogen Mineralization from Animal Manures and its Relation to Organic N Fractions. *J. Integr. Agric.,* 13: 2040-2048.

4. Broadbent, F.E. (1965). Effect of fertilizer nitrogen on the release of soil nitrogen. *Soil Sci. Soc. Am. Pro.,* 29: 692-696.

5. Bremner, J.M. (1965). Organic forms of nitrogen. In: *Methods of Soil Analysis,* Black C.A. *et al.,* ed: *Am. Soc. Agron.* USA 9, Part **2**, 1238-1253.

6. Sihag, D., Singh, J.P., Mehla, D.S. and Bhardwaj, K.K. (2005). Effect of integrated use of inorganic fertilizers and organic materials on the distribution of different forms of nitrogen and phosphorus in soil. *J. Indian Soc. Soil Sci.,* 53(1): 80-84.

7. Abbasi, M.K., Hina, M., Khalique, A. and Khan, S.R. (2007). Mineralization of three organic Manures used as Nitrogen source in a soil Incubated under Laboratory Conditions. *Commun. Soil Sci. Plant Anal.,* 38(13-14): 1691-1711.

8. Qian-Ru, H., Feng, H., Shan, H., Hui-Xin, L., Ying-Hong, Y., Gen-Xing, P. and Wei-Jian, Z. (2009). Effect of Long-Term Fertilization on Organic Carbon and Nitrogen in a Subtropical Paddy Soil. *Pedosphere.* 19(6): 727-734.

9. Tabassum, S., Reddy, K.S., Vaishya, U.K., Singh, M. and Biswas, A.K. (2010). Changes in Organic and Inorganic Forms of Nitrogen in a Typic Haplustert under Soybean-Wheat System due to Conjoint Use of Inorganic Fertilizers and Organic Manures. *J. Ind. Soc. Soil Sci.,* 58(1):76-85.

10. Kaur, J. and Singh, J.P. (2014). Long – term effects of continuous cropping and different nutrient management practices on the distribution of organic nitrogen in soil under ricewheat system. *Plant Soil Environ.,* 60(2): 63-68.

11. Khalil, M.I., Cleemput, O.V., Boeckx, P. and Rosenani, A.B. (2001). Nitrogen transformation and emission of greenhouse gases from three acid soils of humid tropics amended with nitrogen sources and moisture regime. 1. Nitrogen transformations. *Commun. Soil Sci. Plant Anal.,* 32(17-18): 2893-2907.

12. Sajjad, M.H., Azam, F. and Lodhi, A. (2003). Nitrogen transformation in soil amended with different plant residues and their impact on growth of wheat. *Pak. J. Biol. Sciences,* 6(9): 805- 812.

13. Nourbakhsh, F. and Dick, R.P. (2005). Net nitrogen mineralization or immobilization potential in a residue-amended calcareous soil. *Arid Land Res. Manag.,* 19(4): 299-306.

14. Mubarak, A.R., Gali, E.A.M., Mohamed, G., Steffens, D. and Awadelkarim, A.H. (2010). Nitrogen mineralization from five manures as influenced by chemical compositon and soil type. *Commun. Soil Sci. Plant Nutri.,* 41: 1903-1920.

15. Baitilwake, M.A., Salomez, J., Mrema, J.P. and Neve, S.D. (2012). Nitrogen mineralization of two manures as Influenced by Contrasting Application Methods under Laboratory conditions. *Commun. Soil Sci. Plant Anal.,* 43(1-2): 357-367.

16. Helgason, B.L., Larney, F.J., Janzen, H.H. and Olson, B.M. (2007). Nitrogen dynamics in soil amended with composted cattle manure. *Can. J. Soil Sci.,* 87(1): 43-50.

17. Azeez, J.O. and Averbeke, W.V. (2010). Nitrogen mineralization potential of three animal manures applied on a sandy clay loam soil. *Bioresour. Technol.*, 101: 5645-5651.

18. Paul, N., and Saha, D. (2016). Relation between different organic sources and N mineralization in rice-mustard cropping sequence. *J. Indian Chem. Soc.*, 93: 1-9.

19. Flaig, W. (1975). Specific effects of soil organic matter on the potential of soil productivity. 1. Biochemistry of soil organic matter. In: Organic Materials as Fertilizers. Report of the FAO/SIDA, Soils Bull., 27, Rome, pp. 31-69.

20. Flaig, W. and Sochtig, H. (1967). Organische verbindungen als stickstoffquelle fur die ernahrung der pflanzen. *Anales de Edafologia,* 26: 801-828.

21. Schnitzer, M. and Khan, S.U. (1972). Humic substances in the Environment, Marcel Dekker Inc. New York.

22. Filip, Z. and Kubat, J. (2003). Aerobic short - term microbial utilization and degradation of humic acids extracted from soils of long - term field experiments. *Europ. J. Soil Biol.,* 39(4): 175 -182.

23. Nguyen, B.V., Olk, D.C. and Cassman, K.G. (2004b). Nitrogen mineralization from humic acid fractions in rice soils depend on degree of humification. *Soil Sci. Soc. Am. J.,* 68(4): 1278-1284.

24. Satisha, G. C. and Devarajan, L. (2011). Composition and characterization of humic substances extracted from effluent-based pressmud composts. *Agropedology,* 21(I): 8-17.

25. Devi, N.S. and Saha, D. (2017). Influence of Organic Matter Vis-àVis Humic Acid on the Transformation of Inorganic and Organic Forms of Nitrogen in a Typic Haplustept Soil. *Commun. Soil Sci. Plant Anal.*, 48(9): 1042-1051

26. Govindasamy, R. and Chandersakaran, S. (1992). Effect of humic acids on the growth, yield and nutrient content of sugar cane. *Sci. Total Environ.*, 117: 575-581

27. Sharif, M., Khattak, R.A. and Sarir. M.S. (2002). Effect of different levels of lignitic coal derived humic acid on growth of maize plants. *Commun. Soil Sci. Plant Anal.,* 33(19-20): 3567- 3580.

28. Senn, T.L and Kingman, A.R. (1973). A review of Humus and Humic Acids. Horticulture Dept. Research Series No. 145, S.C. *Agric. Exp. Station*, Clemson, South Carolina, USA.

29. Helgason, B.L., Larney, F.J., Janzen, H.H. and Olson, B.M. (2007). Nitrogen dynamics in soil amended with composted cattle manure. *Can. J. Soil Sci.*, 87(1): 43-50.

30. Ahmad, R., Khalid, A., Arshad, M., Zahir, Z. A., Mahmood, T. (2008). Effect of compost enriched with N and L tryptophan on soil and maize. *Agron. Sustain. Dev.*, 28: 299-305

31. Bird, J.A.; Kessel, C.V.; Horwath, W.R. and Van, K.C. (2002). Nitrogen dynamics in humic fractions under alternative straw management in temperate rice. *Soil Sci. Soc. Am. J.*, 66(2): 478 -488.
32. Mohanty, M., Reddy, K.S., Probert, M.E., Dalal, R.C., Rao, A.S. and Menzies, N.W. (2011). Modeling N mineralization from green manure and farmyard manure a laboratory incubation study. *Ecol. Model.*, 222: 719-726.
33. Sen, T.K., Baruah, U., Sarkar, D., Maji, A.K. and Patil, V.P. (2006). Soil Series of Manipur Publication, 134 NBSS and LUP, Nagpur.
34. Stevenson, F.J. (1996). Nitrogen Organic forms. In: Methods of Soil Analysis. Part 3. Chemical methods (Sparks D.L. *et al.* (eds.). Soil Sci. Soc. Am., Book Series No. 5, Madison, WI, USA, pp. 1185-1200.
35. Jackson, M.L. (1973). Soil Chemical Analysis, Prentice Hall of India Pvt. Ltd., New Delhi.
36. Saha, A.K. (2008). Methods of Physical and Chemical Analysis of Soil. Kalyani Publishers, New Delhi.
37. Bremner, J.M. and Keeney, D.R. (1966). Determination of exchangeable ammonia, nitrate and nitrite by extraction distillation methods. *Soil Sci. Soc. Am. Proc.*, 30: 577-587.

Chapter 8

Effect of Nitrogen and Molybdenum on Soil Properties and Yield of Pea (*Pisum sativum* L.)

Ch. Karuna Chanu

ABSTRACT

An experiment entitled "Effect of Nitrogen and Molybdenum on Soil Properties and Yield of Pea (Pisum sativum L.)" was carried out at the research field of Central Agricultural University, Iroisemba, Imphal, Manipur during the Rabi season of 2019 - 2020. The experiment planned in Factorial Randomized Block Design (FRBD) with three replications of each treatment. The unit plot size of the field was 9 m2 (3 m x 3 m). There were 16 treatments combinations in the experiment consisting of 4 levels of nitrogen (N1, N2, N3, N4) viz., 0, 10, 20, 30 kg/ha and 4 levels of molybdenum (M1, M2, M3, M4) viz., 0, 0.4, 0.8, 1.2 kg/ha. The effect of different level of nitrogen and molybdenum was found positive and significant on plant height, number of branches per plant, pod length, number of pods per plant, number of seeds per pod, green pod yield (t/ha), nutrient content in plants and in post- harvest soil. The maximum green pod yield (t/ha) was attained from N4 (30 kg N/ha) which are at par with N3 (20 kg N/ha) and the highest green pod yield (t/ha) was found from M3 (0.8 kg Mo/ha). The highest N, P, K and S content in plants were observed in N3 (20 kg N/ha) and M3 (0.8 kg Mo/ha) and the highest available nitrogen (kg/ha) was recorded in N4 (30 kg N/ha) and M3 (0.8 kg Mo/ha). While the highest available molybdenum (mg/kg) was obtained in N4 (30 kg N/ha) and M4 (1.2 kg Mo/ha). Interaction effect of nitrogen and molybdenum produced significant effect on plant height, number of branches per plant, pod length, number of pods per plant, number of seeds per pod, green pod yield (t/ha), nutrient content in plants and in post- harvest soil. However, *soil pH, organic carbon, available nitrogen (kg/ha), soil texture (sand, silt, clay per cent) found non- significant in post- harvest soil. The maximum green pod yield was attained in the treatment combination N3M3 kg ha^{-1}(T11) while, the minimum was found in control treatment. The highest N, P, K and S content in plants were recorded in the treatment combination N3M3 kg ha^{-1}(T11). The highest available molybdenum content in post- harvest soil was obtained from treatment combination N4M4 kg ha^{-1}(T16) and the highest available nitrogen (kg/ha) were found from N4M3 kg ha^{-1}(T15).The lowest N, P, K, S in plants and available nitrogen (kg/ha) and available molybdenum (mg/kg) in post- harvest soil were attained in the control treatment. From the results of the present investigation, application of N @ 20 kg N/ha and Mo @ 0.8 kg Mo/ha was the most suitable doses for better crop yield.kg N/ha and Mo @ 0.8 kg Mo/ha was the most suitable doses for better crop yield.*

Introduction

Garden pea (*Pisum sativum* L.) is one of the most widely spread legume crops. It belongs to the family of Leguminosae and subfamily of Papilionoideae. The crop is well adapted to cold climate and can self-pollinate itself and can also be grown in tropical countries in the winter season. *Pisum sativum* is the second most important leguminous crop in the world (Pawar *et al.*, 2017). In India it is cultivated as a cold season crop in Northern India plains and as a summer crop in hilly regions. It is believed to have originated from Ethiopia, part of Europe and Asia as reported by Choudhary (1967). Garden pea occupies an approximated area of 540 thousand hectares having a production of 5427 thousand tonnes of grain throughout India (Anonymous, 2017). It is largely cultivated in Jaipur, Baran, Bundi, Kota and Bharatpur districts of Rajasthan.

The garden pea is mostly grown for its green seeds and its sweet taste, so it can be eaten raw without being cooked. This leguminous crop contains a high percentage of digestible protein (7.2g), carbohydrates (15.8g), Vitamin A (139 I.U.), Vitamin C (9 mg), magnesium (34 mg) and phosphorus (139 mg) per 100g of the edible portion (Gopalan, 2007). Generally, garden pea has white flowers and round or wrinkled seeds. The seeds are either green or yellow in colour. This crop is usually harvested at young and succulent stage, and can be shelled and eaten directly; whereas the pods are inedible because of it parchment layer on the inner surface of pods and fibres along the suture lines. The crop and its by-products can also be used as fodders. It can be used as green manure as well (Makasheva.1983). The biomass can be utilised as cattle feeds and also incorporated into soil for supplementing nitrogen for the next crops and to increase organic matter of the soil.

Garden pea has the ability to replenish soil fertility due to their unique ability of symbiotic nitrogen fixation (Rana *et al.*, 1998). As a nitrogen-fixing crop having high assimilating capacity in the roots, it uses the chemical compounds having low solubility and low accessibility to cereals from the cultivated soil layer or from deeper layers. As a preceding crop, it can facilitate an increase in the efficiency of organic matter utilisation by subsequent crops like grain and cash crops. Cultivation of this crop is a highly profitable business and also very attractive to the farmers due to its short durability. Nitrogen is one of the major elements for growth and development of a crop plants (Tanaka *et al.,* 1984). It is an essential element during synthesis of chlorophyll, enzymes and proteins in plants. It is also a component of amino acid-the building blocks of proteins. Nitrogen plays a key role in various physiological processes.

It imparts dark green colour in plants, initiates rapid early growth and improves the capacity to fix the atmospheric nitrogen symbiotically. It promotes the growth and development of leaves, stem and other vegetative parts. It also increases the protein content in harvested peas. Nitrogen utilisation efficiency depends on the soil-N content and method of its application. Nitrogen application to leguminous crops at lower doses in the initial stage is essential for vigorous start. The capacity for nitrogen fixation was reported to be ranging between 64- 121kg/ha (Rennie and Kemp, 1983). Sun and Liu (2000) reported that pea crop having enough formation of nodules can fix approximately 75 kg N/ha. The amount of fixed nitrogen varies with the type of cultivation method of pea, the soil type, the presence of indigenous available nitrogen in the soil and other environmental conditions. Deficit of nitrogen would result in poor growth and stunting of plants (Makasheva, 1983) and consequently reduction to crop yields (Machler *et al.,* 1988; Radin *et al.,* 1988).

Molybdenum (Mo) is an essential micronutrient with an important role in nitrogen (N) metabolism and protein synthesis in plants. Mo acts as a co-factor for nitrogenase enzymes that catalyses the redox reaction to convert the elemental nitrogen (N) into ammonium (NH_4^+) ions during symbiotic nitrogen fixation (Mendel and Hänsch 2002) and nitrate reductase enzymes, which are required for the assimilation of soil nitrates. As a result, plant N metabolism is intricately related to the Mo concentration in soil, especially for leguminous crops (Mendel and Hänsch 2002). Due to this relationship, Mo-deficient legumes results in an unusual proliferation of nodules, leading to nitrogen (N) deficiency (Marschner 2011).

Molybdenum (Mo) is responsible for the formation of nodule tissue and increase in nitrogen fixation (Lewis, 1980 and Sharma *et al.,* 1988). However, the requirement is so small that the seed of grain legumes can contain sufficient amount of molybdenum for the growth of one generation of plant (Harris *et al.,* 1965). It is indispensable for many plant species especially in root nodule legumes because it is directly involved in nitrogen fixation enzyme nitrogenase and nitrogen reduced enzyme nitrate reductase. Without adequate quantities of molybdenum (Mo), nitrogen fixation would not occur and microbial activities are also depressed. Legumes and pulses can produce active nodules only when soils are properly supplied with molybdenum (Ahmed, 1982). Furthermore, Mo availability is also reduced in acidic soil. Malavolta (1949) reported the benefits of molybdenum on the growth of garden peas, which are grown in culture solution containing nitrate nitrogen. Arnon and Stout (1939) first demonstrated this requirement of molybdenum for plant growth using tomato grown hydroponically. Molybdenum deficient plants usually exhibit poor growth and low contents of chlorophyll and ascorbic acid (Marschner 1995, Liu 2002). The direct effect of molybdenum deficiency is on nitrogen metabolism. Hence, Molybdenum has taken an important place in the list of trace elements for plant growth.

In view of these facts, the present experiment have been undertaken with the below objectives-

- To study the effect of different levels of nitrogen and molybdenum application on soil properties, crop growth and yield.
- To study the interaction effect of nitrogen and molybdenum on soil properties, crop growth and yield.
- To identify the suitable doses of nitrogen and molybdenum for better crop yield.

Materials and Methods

Description of Study Site

The experiment was conducted at the research field in College of Agriculture, Central Agricultural University, Iroishemba, Imphal, Manipur during the Rabi season of 2019-2020. The mean maximum and minimum temperature recorded during the cropping season was 23.25 and 8.4°C, respectively. The rainfall recorded was 3.3 mm. The average relative humidity in 700h was 88.5 per cent and in 1300h was 43 per cent. Composite soil samples were collected at random from the depth of 0-15cm from the experimental field before starting the experiment. A homogenous composite sample was then prepared for mechanical and chemical analysis. Result of the analysis and method used for determination of the soil properties are presented in Table 8.1.

Table 8.1. Physio-chemical Parameters of the Initial Soil of the Experiment

Parameters	*Value*	*Status*	*Method followed*
A. Physical properties			
1. Soil texture			
i) Sand (per cent)	18.3	Clay	Hydrometer method (Bouyoucos, 1962)
ii) Silt (per cent)	30		
iii) Clay (per cent)	51.7		
B. Chemical properties			
1. pH	5.4	Acidic	1:2.5 soil: water suspension (Jackson, 1973)
2. Organic carbon (per cent)	1.32	High	Walkley and Black titration (Walkley and Black, 1934)
3. Available N (kg/ha)	250.88	Medium	Alkaline Permanganate (Subbiah and Asija, 1956)
4. Available P_2O_5 (kg/ha)	18.12	Medium	Bray's method (Bray and Kurtz, 1945)

Parameters	*Value*	*Status*	*Method followed*
5. Available K_2O (kg/ha)	238.42	Medium	Flame photometer (Jackson, 1973)
6. Available Sulphur (mg/kg)	17.66	High	Turbidimetric method (Williams and Steinbergs,1959)
7. Available Molybdenum (mg/kg)	0.025	Low	Ammonium oxalate (pH-3.3) (Griggs,1953)

Experimental Treatments and Design

a) Treatment Details

The experiment was undertaken to study the effect of 4 levels of nitrogen and 4 levels of molybdenum on the growth and yield of pea. The study comprises the following treatment.

Nitrogen level: 4

1) N1: 0 (Control)
2) N2: 10 kg N/ha
3) N3: 20 kg N/ha
4) N4: 30 kg N/ha

Molybdenum level: 4

1) M1: 0 (Control)
2) M2: 0.4 kg Mo/ha
3) M3: 0.8 kg Mo/ha
4) M4: 1.2 kg Mo/ha

There were 16 treatment combinations of nitrogen and molybdenum levels used in the experiment as followings:

- ☆ N1M1 - Control
- ☆ N1M2 - 0 kg N/ha + 0.4 kg Mo/ha
- ☆ N1M3 - 0 kg N/ha + 0.8 kg Mo/ha
- ☆ N1M4 - 0 kg N/ha + 1.2 kg Mo/ha
- ☆ N2M1 - 10 kg N/ha + 0 kg Mo/ha
- ☆ N2M2 - 10 kg N/ha + 0.4 kg Mo/ha
- ☆ N2M3 - 10 kg N/ha + 0.8 kg Mo/ha
- ☆ N2M4 - 10 kg N/ha + 1.2 kg Mo/ha
- ☆ N3M1 - 20 kg N/ha + 0 kg Mo/ha
- ☆ N3M2 - 20 kg N/ha + 0.4 kg Mo/ha
- ☆ N3M3 - 20 kg N/ha + 0.8 kg Mo/ha
- ☆ N3M4 - 20 kg N/ha + 1.2 kg Mo/ha
- ☆ N4M1 - 30 kg N/ha + 0 kg Mo/ha
- ☆ N4M2 - 30 kg N/ha + 0.4 kg Mo/ha
- ☆ N4M3 - 30 kg N/ha + 0.8 kg Mo/ha
- ☆ N4M4 - 30 kg N/ha + 1.2 kg Mo/ha

b) Layout and Design

The experiment was laid out in Factorial randomized block design with three replications. The experimental plot was first divided into six blocks. Each block consisted of 8 units of plots. Different combination of N and Mo were assigned randomly to each block as per design of the experiment. Cultivar Arkel was selected for the trial, It is an exotic variety from France. The plant is dwarf but the growth is vigorous and may grow upto 45 cm. Flowers are white and are borne in singly double on 6 to 7th node. Pods are dark green, 8.5 cm long, incurved towards the sutures and pointed distal end with 8 to10 well filled, wrinkled seeds. It is suitable for both fresh market and dehydration. All recommended agronomic practices and disease control measures were adopted for healthy and disease free of growth of crop during the cultivation period. Green pods were harvested at proper edible maturity stage when young tender filled peas changing in colour from dark to light green.

Soil Sampling and Plant Analysis

Soil samples were collected at 0-15 cm soil depths from five locations of each plot of the experimental field after the harvesting of crop. A homogenous composite sample was then prepared for the mechanical and chemical analysis.

Soil texture of the soil was done by hydrometer method (Bouyoucos,1962). The textural class was determined using Marshall's Triangular Co-ordinate as designated by USDA (1951).

The glass electrode systronic pH meter was used to determine the pH of the soil samples of each plot. 1:2.5 Soil:Water suspension method as described by Jackson(1973).

Organic carbon in soil was determined by Walkley and Black' method (Walkley and Black, 1934). The underlying principle is to oxidize the organic carbon present in organic matter with potassium dichromate in presence of sulphuric acid and to titrate the residual K2Cr2O7 solution with standard ferrous ammonium sulphate. The result was expressed in percentage.

Available nitrogen (kg/ha) in soil of each plot was determined by Alkaline permanganate method (Subbiah and Asija, 1956) in which the soil is mixed with excess of alkaline $KMnO_4$ solution and distilled. Organic matter present in the soil is oxidised by nascent oxygen liberated by KMnO4 in the presence of NaOH and thus ammonia released is distilled and is absorbed in a known volume of standard acid, excess of which is back titrated with standard alkali using methyl red indicator.

Available phosphorus (kg/ha) was estimated by Bray's and Kurtz No.1 method (Bray and Kurtz, 1945). The underlying principle is that the combination of HCl and NH4F extracts acid soluble forms of P largely calcium phosphate and a portion of Al and Fe phosphates. The fluoride ions complex trivalent Al and Fe in acid solution and thereby release P held by Al and Fe.

Available potassium (kg/ha) content in the soil samples were analysed by adopting flame photometer method as by Jackson (1973).

Available Sulphur (mg/kg) in soil was determined by turbidimetric method as described by Williams and Steinbergs (1959).

Available Molybdenum (mg/kg) was determined by Griggs reagent method given by Grigg (1953). Ammonium oxalate method is the most popular method for extraction of Mo from soil. Colour intensity was read at 680 nm using spectrophotometer.

Five plants were randomly selected and tagged from each treatment under each replication excluding the border plants. Observations were recorded from the tagged plants for biometric attributes like Plant height (Cm), Number of branches per plant, Pod length (cm), Number of pods per plant, Number of seeds per pod and Green pod yield (t/ha).

Plant samples from recent fully developed leaflets at first blooming were collected, decontaminated and washed with distilled water, dried in oven at 80°C, ground to a 20 mesh size and stored at 4°C for analysis. Samples were analysed for Nitrogen (Kjeldahl Method), Phosphorus (Jackson, 1973). Potassium (Jackson, 1973), Sulphur content Williams and Steinbergs (1959).

Statistical Analysis

The data obtained from the experiments will be analyzed with the help of Analysis of variance technique after suitable transformation (if necessary) for comparison of the treatment effects and proper interpretation of the result.

Brief Findings

Based on the findings of the experiment the individual and interaction effect of nitrogen (N) and molybdenum (Mo) on soil properties, crop growth and yield were significantly influenced. Data on plant height and number of branches per plant were taken at 30, 45, 60, 75 DAS and at maturity. Application of nitrogen level N3 (20 kg N/ha) obtained the highest plant height at 30 and 45 DAS and the maximum plant height at 60 DAS, 75 DAS and at maturity were recorded from the nitrogen level N4 (30 kg N/ha). For number of branches per plant, application of nitrogen level N4 (30 kg N/ha) recorded the maximum number at 30 DAS, 60 DAS and at maturity whereas, in case of 45 DAS and 75 DAS, the highest number of branches per plant were obtained from the nitrogen level N3 (20 kg N/ha). The lowest plant height and number of branches per plant recorded in 30, 45, 60, 75 DAS and at maturity were found from the nitrogen level N1 (control). Application of nitrogen level N3 (20 kg N/ha) observed the maximum pod length and number of seeds per pod and the highest number of pods per plant and green pod yield were obtained from the nitrogen level N4 (30 kg N/ha). However, the lowest pod length, number of pods per plant, number of seeds per pod, green pod yield were found in the nitrogen level N1 (control).

Different level of molybdenum on crop growth and yield were influenced significantly. Application of molybdenum level M4 (1.2 kg Mo/ha) obtained the highest plant height and highest number of branches per plant at 30, 45, 60, 75 DAS and at maturity. Whereas, the lowest plant height and number of branches per plant at 30, 45, 60, 75 DAS and at maturity were found in the nitrogen level N1 (control). The maximum pod length, maximum number of pods per plant, highest number of seeds per pod and maximum green pod yield were recorded from the plants which was treated with molybdenum level M3 (0.8 kg Mo/ha) and the lowest pod length, minimum number of 70 pods per plant, lowest number of seeds per pod and minimum green pod yield were found from the plants which was treated with molybdenum level M1 (control).

The interaction effect of nitrogen and molybdenum on crop growth and yield was found significant. The treatment combination N4M3 kg ha^{-1}(T15) gave the highest plant height at 30 DAS and the maximum plant height at 45, 60, 75 DAS and at maturity were obtained from the treatment combination N4M4 kg ha^{-1}(T16). For number of branches per plant, the treatment combination N3M4 kg ha^{-1}(T12) produced the highest number of branches per plant at 30, 45, 60, 75 DAS and at maturity. However, the interaction effect of nitrogen and molybdenum on number of branches per plant was non-significant at 30 DAS. While, the lowest plant height and lowest number of branches per plant at 30, 45, 60, 75 DAS and at maturity were found in the control treatment (T1). The treatment combination N3M3 kg ha^{-1}(T11) gave the highest pod length, maximum number of pods per plant, highest number of seeds per pod and maximum green pod yield whereas, the lowest pod length, minimum number of pods per plant, lowest number of seeds per pod and minimum green pod yield were obtained from the control treatment (T1).

Nutrients content in plants was significantly influenced by individual and interaction effect of nitrogen and molybdenum. Application of nitrogen level N3 (20 kg N/ha) produced the highest N, P, K and S content in plants and the lowest N, P, K and S content in plants were obtained from the nitrogen level N1 (control). Molybdenum level M3 (0.8 kg Mo/ha) gave the highest N, P, K and S content in plants while, the lowest N, P, K and S content in plants were obtained from the molybdenum level M1 (control). The treatment combination N3M3 kg ha^{-1}(T11) produced the highest N, P, K and S content in plants whereas, the lowest N, P, K and S content in plants were from the control treatment.

Nutrients content in post-harvest soil was influenced by the application of different levels of nitrogen and molybdenum. Application of nitrogen level N4 (30 kg N/ha) gave the highest soil pH, maximum available nitrogen (kg/ha), maximum available molybdenum (mg/kg) and highest organic carbon (per cent). However, the highest available phosphorus (kg/ha), highest available potassium (kg/ha), highest available sulphur (mg/kg) were obtained in the nitrogen level N1 (control). The lowest soil pH, organic carbon (per cent), available nitrogen (kg/ha), available molybdenum were found in the nitrogen level N1 (control) and the lowest available phosphorus (kg/ha), available potassium (kg/ha), available sulphur (mg/kg) were obtained from the nitrogen level N4 (30 kg N/ha). However, the effect of different level of nitrogen on soil texture (sand, silt and clay per cent) was found non- significant.

Molybdenum level M4 (1.2 kg Mo/ha) produced the highest soil pH, available molybdenum (mg/kg), highest organic carbon (per cent) and the highest available nitrogen (kg/ha) were noted in the molybdenum level M3 (0.8 kg Mo/ha). While, the highest available phosphorus (kg/ha), highest available potassium (kg/ha), highest available sulphur (mg/kg) were obtained in the molybdenum level M1 (control). The lowest soil pH, available molybdenum (mg/kg), organic carbon (per cent), available nitrogen (kg/ha) were noted in the molybdenum level M1 (control) while, the lowest available phosphorus (kg/ha), available potassium (kg/ha), available sulphur (mg/kg) were obtained in the molybdenum level M4 (1.2 kg Mo/ha). Different levels of molybdenum on soil texture (sand, silt and clay per cent) was found non- significant.

The highest soil pH, highest organic carbon (per cent), highest available molybdenum (mg/kg) was obtained from the treatment combination N4M4 kg ha^{-1}(T16) and the treatment combination N4M3 kg ha^{-1}(T15) gave the highest available nitrogen (kg/ha). Whereas, the highest available phosphorus (kg/ha), available potassium (kg/ha), available sulphur (mg/kg) were noted in the control treatment. The lowest soil pH, available nitrogen (kg/ha), organic carbon (per cent), available molybdenum (mg/kg) were obtained in the control treatment and the lowest available phosphorus (kg/ha), available potassium (kg/ha), available sulphur (mg/kg) were noted in the treatment combinations N4M4 kg ha^{-1}(T16). However, the interaction effect of nitrogen and molybdenum on soil pH, soil texture (sand, silt and clay per cent), organic carbon (per cent), available nitrogen (kg/ha), was non-significant.

Conclusion

- From the results of the present investigation, the following conclusion could be drawn:
- Individual effect of nitrogen and molybdenum on crop growth and yield was found positive and significant. The highest yield was found in application of N @ 20 kg N/ha and Mo @ 0.8 kg Mo/ha.
- The interaction effect of nitrogen and molybdenum enhanced crop growth and yield with the treatment of N3M3 kg/ha.
- Individual and interaction effect of nitrogen and molybdenum on nutrient content in plants was found significant. Treatment combination N3M3 kg/ha obtained the highest nutrient content in plants.
- Individual effect of nitrogen and molybdenum on chemical properties of postharvest soil was found significant. However, interaction effect of nitrogen and molybdenum for P, K, S and Mo were recorded significant.

- ✰ Application of N @ 20 kg N/ha and Mo @ 0.8 kg Mo/ha was the most suitable doses for better crop yield.

REFERENCES

1. Pawar, Y., Varma, L.R., Verma, P., Joshi, H.N., More, S.G. and Dabhi, J.S. (2017). Influences of integrated use of organic and inorganic sources of nutrients on growth, flowering and yield of garden pea (Pisum sativum L.) cv. Bonneville. Legume Research, 40(1): 117-124.
2. Choudhary B. (1967) Vegetable (Ist Ed.) National Book Trust, India, 113.
3. Anonymous. (2017) Indian Horticulture Database, National Horticulture Board, Gurgaon.
4. Gopalan, C., Rama, S.B.V. and Balasubramanian, S.C. (2007). Nutritive Value of Indian Foods revised edition. National Institute of Nutrition, Indian Council of Medical Research, Hyderabad, India. P. 50
5. Rana, N.S., Singh, G.V. and Ahlawat, I.P.S. (1998). Effect of nitrogen, rhizobium inoculation and phosphorus on root nodulation, dry matter yield and nutrient uptake in pigeon pea (Cajanas cajan). Indian Journal of Agronomy, 43: 102-106.
6. Tanaka, A.J. Yamaguchi, Miura, S. and Tamaru, H. (1984). Comparison of fertilizer nitrogen efficiency among field crops. Soil Science and Plant Nutrition, 30(2): 199-208.
7. Rennie, R.J. and Kemp, G.A. (1983). N2-Fixation in Field Beans Quantified by 15N Isotope Dilution. II. Effect of Cultivars of Beans 1. Agronomy Journal, 75(4):645-649.
8. Sun, Y.Y. and Liu, G.F. (2000). Pea varieties and cultivation techniques. Golden Shield Press, Beijing, China.
9. Makasheva, R.K.H. (1983). The pea. Oxonion Press Pvt. Ltd. New Delhi. pp. 267.
10. Machler, F., Oberson, A., Grub, A. and Nosberger, J. (1988). Regulation of photosynthesis in nitrogen deficient wheat seedlings. Plant Physiology, 87: 46-49.
11. Radin, J.W., Hartung, W., Kimball, B.A. and Mauney, J.R. (1988). Correlation of stomatal conductance with photosynthetic capacity of cotton only in a CO2 enriches atmosphere mediation by abscisic acid. Plant Physiology, 88: 1058-1062.
12. Mendel, R. R., and Hänsch, R. (2002). Molybdoenzymes and molybdenum cofactor in plants. Journal of experimental botany, 53(375):1689-1698.
13. Marschner, H. (1995). Mineral nutrition of higher plants. second edition. 889 pp. London: Academic Press, Annals of Botany, 78(4): 527–528.
14. Lewis, D.H. 1980. Boron, lignification and the origin of vascular plants infield hypothesis, New phytologist, 84:209 –229.

15. Sharma, M. S.; M. S. Upadhyay and S. S. Tomar. (1988). Water use efficiency of some rainfed crop on vertisol as influenced by soil micronutrient and straw mulching. Indian Journal of Soil Science,33:387-390.
16. Harries, H.B., Parker M. B. and Johnson B.J. (1965). Influenced of molybdenum content of soybean seed and other factors associated with seed source on progeny response to applied molybdenum. Agronomy Journal, 57: 397 –399.
17. Ahmed, S. (1982). Crops response to applied molybdenum in Brahmaputra Flood Plain Soil of Bangladesh. Journal of Soil Science, 18:36 -41.
18. Malavolta, E. (1949). Studies on the nitrogenous nutrition of Garden Pea. Plant Physiology, 29: 98-99.
19. Arnon, D.I. and Stout, P.R. (1939). Molybdenum as an essential element for higher plants. Plant Physiology, 14(3): 599.
20. Liu, P. (2002). Effects of the stress of molybdenum on plants and the interaction between molybdenum and other element. Agri-Environmental Protection, 21: 276–278.
21. Bouyoucos, G.H. (1962).The hydrometer as a new method for the mechanical analysis of soils. Agronomy Journal, 54: 464-465.
22. Jackson, M.L. (1973). Soil Chemical Analysis. Prentice Hall of India Private limited, New Delhi.
23. Walkley, A.J. and Black, I.A. (1934). Estimation of soil organic carbon by the chromic acid titration method. Soil Science, 37: 29-38.
24. Subbiah, B.V. and Asija, G.L. (1956). A rapid procedure for estimation of available nitrogen in soils. Current Science, 25: 259-260.
25. Bray, R.H. and Kurtz, L.T. (1945). Determination of total organic and available forms of phosphorus in soils. Soil Science, 59: 39-45.
26. Williams, C. H., and Steinbergs, A. (1959). Soil sulphur fractions as chemical indices of available sulphur in some Australian soils. Australian Journal of Agricultural Research, 10(3): 340-352.
27. Grigg, J. L. (1953). Determination of the available molybdenum of soils. New Zealand Journal of Science and Technology, 34: 405-414.

Section D

Nutrient Management in Vegetables/Citrus Fruits

Chapter 9

Effect of Integrated Nutrient Management on Soil Fertility and Productivity in Potato in Acid Soil

Dorjee Tsering Bapu

ABSTRACT

A field experiment on "Effect of integrated nutrient management on soil fertility and productivity in potato (Solanum tuberosom L.) in acid soil" was conducted during the Rabi season of 2015-16 at the research farm of College of Agriculture, CAU, Imphal, Manipur. The experiment was laid out in Randomized block Design with 15 treatments and replicated thrice. The treatment constituted of synthetic fertilizers, vermicompost, biofertilizers at different rates viz., T1=100 per cent RDN V.C, T2=100 per cent RD NPK, T3=50 per cent RD NPK+50 per cent RDN V.C, T4=50 per cent RD NPK+50 per cent RDN V.C+ Azotobacter, T5=50 per cent RD NPK+50 per cent RDN V.C +PSB, T6=50 per cent RD NPK+50 per cent RDN V.C + PSB and Azotobacter, T7=75 per cent RD NPK+25 per cent RDN V.C, T8=75 per cent RD NPK+25 per cent RDN V.C+ Azotobacter, T9=75 per cent RD NPK+25 per cent RDN V.C +PSB, T10=75 per cent RD NPK+25 per cent RDN V.C +PSB and Azotobacter, T11=25 per cent RD NPK+75 per cent RDN V.C, T12=25 per cent RD NPK+75 per cent RDN V.C+ Azotobacter, T13=25 per cent RD NPK+75 per cent RDN V.C+PSB, T14=25 per cent RD NPK+75 per cent RDN V.C +PSB and Azotobacter, T15= CONTROL.

Plot dimensions were 4×2m with a buffer zone of 0.5 m between all the plots. A spacing of 20×45 cm between the plants was maintained with a total of 45 plots. The soil of the experimental site was clay in texture with pH 5.3, CEC 12.8 and 2.2 per cent organic carbon, medium available N (338.6 kg/ha), P_2O_5 (22.02 kg/ha) and high K_2O (320 kg/ha) in 0-15 cm *soil depth and pH 5, CEC 10.4 and 1.8 per cent organic carbon, medium available N (312.2 kg/ha), low P_2O_5 (18 kg/ha) and medium K_2O (290 kg/ha) in 15-30 cm soil depth. From the investigation it was found that all the growth character such as plant height, no of leaves, no of stems, recorded at 30,60 and 90 days was recorded higher in treatment constituting 75 per cent RD NPK+25 per cent RDN V.C +PSB and Azotobacter and was significant to most of the treatments including control. Improvement in available nutrients content i.e, NPK and other Physico-chemical properties of soil like soil pH, organic carbon, CEC, was seen in most of the treatments from the initial status except in control. The Maximum improvement in fertility was noticed with application of 100 per cent RDN through vermicompost or treatment which content 75 per cent of vermicompost. The yield both in grade wise and total yield per plot was recorded highest in treatment T10 with 75 per cent RD NPK+25 per cent RDN V.C +PSB and Azotobacter (19.63 t/ha) along with maximum net return (193091.5) and B: C ratio (1.9). Thus it can be concluded that conjoint use of 75 per cent RD NPK+25 per cent RDN V.C along with biofertilizers not only reduced the use of chemical fertilizers but also enriched the fertility status of soil for sustainable production of potato.*

Introduction

Potato (*Solanum tuberosum* L.) is one of the important crops of the world and is consumed by people across the globe. It is originated from Peru and Bolivia. It is being cultivated in India for the last 300 years. In India potato is an important vegetable crop that occupies a unique position among food crops, for vegetable purpose. Considering its usage with almost all other vegetables used for consumption after cooking it has become one of the most popular crop in the country. Potato is a rich source of vitamins, proteins, carbohydrates, enzymes and other substances necessary for human nutrition. It is also used for several industrial purposes such as for production of starch, alcohol, dextrin, glucose, dyes *etc.*

The post green revolution has witnessed substantial increase in production of food but at the cost of soil health and environmental pollution. Potato crop is grown in an area of 19.3 million hectares in 150 different countries of the world with a total production of 308 million tonnes globally.

In India, potato is cultivated in an area of about 20.32 lakh hectare with a total production of about 46.61 million tonnes. Uttar Pradesh, West Bengal, Bihar and Punjab are the leading states involved in cultivation of potato. The productivity of potato is very low in North Eastern hill (NEH) region of India (8.64 t/ha) except in Tripura (17.3t/ha) due to poor management practices (Burman *et al.,* 2007).

Potato is well adapted to temperate regions but it can also be cultivated successfully under sub-tropical areas. It thrives best under short day condition coupled with abundant sunshine and cool nights. Potato can be grown on a wide range of soils ranging from sandy loam to heavy clay, however well drained sandy loam soils are well suited for higher yield. It grows well in a pH ranging from 4.8 to 7.0.

With the strongly undulated topographies, Manipur (22,327 sq. km) has different types of climates such as sub-tropical in the central valley and semi-temperate forms of climate in the hills. The nature of the soil also varies from clay to loam in most part of the valley whereas in the hill areas soils are mostly lateritic in nature with pH value of 4.7 to 6.8.

In Manipur, potato crop is cultivated in the valley during winter season where irrigation can be provided for optimum yield production. It is grown in an area of 5,500 ha producing 55,000 metric tonnes (as on 1990) with an average yield of 100 q/ha which is rather low in terms of per ha yield as compared to all-India average,(Meitei and Singh, 2001).

A number of diverse organic sources are available for use in agriculture in the NEH region due to its favorable agro-climatic conditions. These sources can reduce the exploitation of soil nutrient and improve overall soil productivity in terms of improved Physico-chemical and biological conditions of the soil.

Ideal soils are rarely found in nature. Organic manures not only supply the plant nutrient but also improve soil health. Diminishing soil fertility is one of the bottle necks to sustain agricultural production and productivity.

The integrated nutrient management (INM) systems envisage use of organic manures along with chemical fertilizers, Bio-fertilizers like solubilizing bacteria (PSB)/Azotobacter may be useful for improving P and N nutrition in potato. Integrated nutrient management is the best approach in which all the possible source of plant nutrients are applied based on economic consideration. The combined use of organic and inorganic source of plant nutrient not only pushes the production and profitability of field crops but also helps in maintaining the permanent fertility status of soil.

In the recent past the declining or stagnating yield trend in India has been attributed to multiple nutrient deficiencies and imbalance of nutrients. Therefore, the most widely recognized alternative is integrated nutrient management. Safe guarding soil health and improving soil fertility in fields is the need of the hour. Since, no single source is capable of supplying the required amount of plant nutrients, integrated use of all sources of plant nutrients is a must to supply balanced nutrition to the crops. Nevertheless, a lot of integrated approaches have been made for pushing up the production of vegetables crops. However, meagre work has done on potato crop.

Keeping this in view, the present investigation entitled "*Effect Of Integrated Nutrient Management on Soil Fertility and Productivity in Potato (Solanum Tuberosum L.) in Acid Soil*" was taken up during Rabi *season* 2015-16 at the Research Farm of the College of Agriculture, Central Agricultural University, Imphal with the following objectives:

1. To identify the effect of Integrated Nutrient Management on Physico-chemical properties of soil.

2. To study the effect of Integrated Nutrient Management on yield of tuber crop and
3. To study the effect of Integrated Nutrient Management on economics of potato production

Materials and Methods

Description of Study Site

Field experiment was conducted during the *Rabi* season of 2015 at the research farm of College of Agriculture, CAU, Imphal. The area is 784 m above mean sea level and experienced an average of 1881 mm of precipitation annually with an average day temperature of 20°C. All analysis work was done at the laboratories of Department of Soil Science and Agricultural Chemistry, CAU, Imphal.

Soil Sampling and Analysis

Before planting, Soil samples were taken from 0-15, 15-30 Cm depth for analysis of Physical parameters (Initial) and Chemical parameters (Initial, 60 days and Postharvest (120) interval. The Soil collected was thoroughly mixed and analysed.

Physico Chemical Properties of the Soil before Growing the Crop

Physical Parameters

Soil property	*Value*		*Method of Analysis*
a) (0-15 cm soil depth)			
Sand per cent	6.6	Clay soil	Hydrometer method. (Bouyoucos 1951)
Silt per cent	12.5		
Clay per cent	81.36		
b) (15-30 cm soil depth)			
Sand per cent	5.8	Clay soil	Hydrometer method. (Bouyoucos 1951)
Silt per cent	10.2		
Clay per cent	76		
Soil property	0-15 cm (depth)	15-30 cm (depth)	
Bulk density	1.38 g/cc	1.42 g/cc	
Moisture content (per cent)	25.13 per cent	22.3 per cent	

Chemical Properties

Soil Property	*Value*	*Interpretation*	*Method of Analysis*
a) (0-15 cm soil depth)			
Organic carbon	2.2	High	Wet digestion method Walkley and Black, 1934
pH	5.3	Acidic	1:2.5 soil: water suspension Jackson,1973
Available nitrogen (N) Kg/ ha.	338.6	Medium	Alkaline permanganate method Subbiah and Asija,1956
Available phosphorus (P) Kg/ha	22.02	Medium	Brays method Bray and Kurtz,1945
Available potassium (K) Kg/ha	320	High	Neutral normal ammonium acetate method, (Jackson, 1973)
CEC	12.8	Medium	Standard alkali Borah *et al.*, 1987
b) (15-30 cm soil depth)			
Organic carbon	1.8	High	Wet digestion method Walkley and Black, 1934
pH	5	Acidic	1:2.5 soil: water suspension (Jackson,1973)
Available nitrogen (N) Kg/ ha.	312.2	Medium	Alkaline permanganate method Subbiah and Asija,1956
Available phosphorus (p) Kg/ha	18	Low	Brays method Bray and Kurtz,1945
Available potassium (k) Kg/ ha	290	Medium	Neutral normal ammonium acetate method, (Jackson, 1973)
CEC	10.4	Medium	Standard alkali Borah *et al.*, 1987

Experimental Treatments and Design

Experimental design	:	RBD
Number of replications	:	3
Number of treatments	:	15
Plot size	:	4m×2m
Borders between replications:	:	0.5m
Borders between plots	:	0.5m

Net experimental areas	:	360 m^2 16
Gross experimental areas	:	546 m^2
Spacing	:	20×45 cm
Total number of plots	:	45
Variety	:	*Kufri jyoti*

Treatment Details

The details of the treatment with their notations were as follows

Notation	*Treatments*
T1	100 per cent RDN V.C
T2	100 per cent RD NPK
T3	50 per cent RD NPK+50 per cent RDN V.C.
T4	50 per cent RD NPK+50 per cent RDN V.C+ Azotobacter
T5	50 per cent RD NPK+50 per cent RDN V.C +PSB.
T6	50 per cent RD NPK+50 per cent RDN V.C +PSB and Azotobacter.
T7	75 per cent RD NPK+25 per cent RDN V.C.
T8	75 per cent RD NPK+25 per cent RDN V.C+ Azotobacter
T9	75 per cent RD NPK+25 per cent RDN V.C +PSB.
T10	75 per cent RD NPK+25 per cent RDN V.C +PSB and Azotobacter.
T11	25 per cent RD NPK+75 per cent RDN V.C.
T12	25 per cent RD NPK+75 per cent RDN V.C+ Azotobacter
T13	25 per cent RD NPK+75 per cent RDN V.C +PSB.
T14	25 per cent RD NPK+75 per cent RDN V.C +PSB and Azotobacter.
T15	CONTROL

Land Preparation

The field was ploughed using tractor and manual labour before planting and brought to a fine tilth.

Application of Fertilizer

Dose and Source of Fertilizer

The crop was fertilized with a uniform dose of nitrogen 120 kg/ha, phosphorus of 120 kg/ha and potassium of 60 kg/ha, respectively by Urea, SSP and MOP. As per the gross plot size, the actual quantity of fertilizer were computed and applied to the field.

Time and Method of Fertilization

Half dose of N and full dose of P and K (urea, SSP, MOP, respectively) was applied as a basal dose and Half of N dose at earthing up at (35 days). Vermicompost was applied before final land preparation (As per treatment) and biofertilzers was applied through soil treatment (5 kg/ha).

Tuber Plantation

Tuber planting was done in November 12, 2015 using 50 kg of potato in 360 m2 net experimental area at a spacing of 45×20 cm at a depth of 10 cm.

Irrigation

The crop was irrigated for five times at 15 days interval with effect from 10 days after planting including the irrigation just after planting and kept the soil moist throughout the growing period.

Earthing Up

Earthing up was done at 35 DAS when the plants attained 3 to 4 leaved stage.

Harvesting of Tubers

Dehaulming was done manually when the lower leaves of the crop started yellowing and after 120 days it was harvested. The harvested tubers were left to the respective plots for curing. Then the tubers were collected separately for further observation. 18

Observations Recorded

From each plot, 5 plants were randomly selected and were tagged to be sample plants excluding the plant situated in the border rows. All the observations recorded regarding to growth and yield attributes of potato were done from these tagged plants (sample plants). The growth parameters were recorded at 30, 60 and 90 days after planting while yield attribute were taken at the time of harvesting.

Emergence of Seedling

The number of emergence of seedlings of the crop was recorded at 5 days interval with effect from 20 days after sowing till 30 DAS and it was converted to percentage.

Morphological Observation

The individual plant observation for the following characters was taken at a interval of 30, 60, 90 days.

Number of Stems per Plant

Total number of stems including main stem and secondary stems per plant were counted from the sample plants at 30, 60 and 90 DAS and their mean was recorded as number of stems per plant.

Plant Height (cm)

The height of the plant in centimetre was determined from the sample plants by measuring the height from the ground level to the tip of the longest stem or main stem and their mean was computed as plant height.

Number of Leaves per Plant

All the active compound leaves were counted from the tagged plants and average has been taken to be number of leaves per plant.

Leaf Area (cm^2)

Leaves from the tagged plants was collected and leaf area was measured by the formula,

$$Y = (3.81 + 0.393X)\ N$$

where,

Y= Leaf area (cm^2)

X= the products of length and breadth of leaves

N= Total number of leaves/plant. 19

Tuber Yield (tonnes/ha)

Total tuber yield for individual plot were harvested and was recorded and converted into tonnes per hectare and grading was done into A, B and C grades.

Soil Analysis

Physical Properties

Particle Size Analysis

Forty grams of soil samples was taken in a beaker and about 300 ml of distilled water was added. It was then placed on a hot plate and 5 ml hydrogen peroxide was added. Addition of hydrogen peroxide continued until all the organic matter were oxidized. Then 20 ml of 5 per cent sodium hexametaphosphate solution was added to the organic matter free soil. The soil was then allowed to soak it for 15 minute. Then the soil suspension was stirred with the help of an electric stirrer for 20 minute. Silt, sand, and clay percentage were determined by Bouyoucos hydrometer method (Bouyoucosm 1927).

Soil Moisture

Soil moisture was measured by gravimetric method by placing the weighed soil sample in an oven at 105°C and dried to constant weight. The weight difference is considered to be water present soil sample.

Bulk Density

Bulk density of the soil sample was measured by core method.

Chemical Properties

Soil Reaction (pH)

The pH of the soil sample was determined in soil water suspension (1:2) by using glass electrode systronic pH meter.

Organic Carbon

Organic carbon present in the soil samples was oxidized to carbon dioxide in the presence of potassium dichromate and concentrated sulfuric acid and was determined by the Walkley and black (1934) rapid titration method. 20

Available Nitrogen

Available nitrogen of the soil was determined by alkaline permanganate method as described by Jackson (1973).

Available Phosphorous

Available phosphorus contents of the soils were determined spectrophotometrically by chloro stannos-reduced molybdophosphoric blue colour method in hydrochloric acid systems using 0.03 N NH4F in 0.0025 N HCL as extractant (Bray and kurtz) as described by Jackson (1973).

Available Potassium

Available potassium of the soil was determined by flame photometer method as described by Jackson (1973).

CEC (Cation Exchange Capacity)

Cation exchange capacity was determined by 1N ammonium acetate extract.

Economic Analysis

Utility of adopting different practices was compared by using the following economic parameters:

Gross returns = Tuber yield x local market price

Net returns = Gross returns – Cost of cultivation

Benefit cost ratio = Net returns (Rs./ha)/Cost of cultivation (Rs./ha)

Method of Statistical Analysis

The analysis of variance method as suggested by Gomez and Gomez (1984) was followed to statistically analysis the various data recorded and interpretation of the results. The significant of the different sources of variation was tested with the help of F test at 5 per cent level of probability.

Brief Findings

The experiment for the present investigation was laid out in Randomised Block

Design with 15 treatments and three replications. Plot dimensions of 4×2m with a buffer zone of 0.5 m between all plots. A spacing of 45×20 cm between the plants was maintained with a total of 45 plots. The area is 784 m above mean sea level and experienced an average of 1881 mm of precipitation annually with an average day temperature of 20°C. All the analysis was done at the laboratories of Department of Soil Science and Agricultural Chemistry, CAU, Imphal, Manipur.

The soil of the experimental site was clay in texture with pH 5.3, CEC 12.8 and 2.2 per cent organic carbon, medium available N (338.6 kg/ha), P_2O_5(22.02 kg/ha) and high K_2O (320 kg/ha) in 0-15 cm soil depth and pH 5, CEC 10.4 and 1.8 per cent organic carbon, medium available N (312.2 kg/ha), low P_2O_5 (18 kg/ha) and medium K_2O (290 kg/ha) in 15-30 cm soil depth The crop was fertilized with a uniform dose of nitrogen 120 kg/ha, phosphorus of 120 kg/ha and potassium of 60 kg/ha, respectively by Urea, SSP and MOP. As per the gross plot size and treatment dosage, the actual quantity of fertilizer were computed and applied to the field. Half dose of N and full dose of P and K (urea, SSP, MOP, respectively) was applied as a basal dose and Half of N dose at earthing up at (35 days). Vermicompost was applied before final land preparation (As per treatment) and biofertilzers was applied through soil treatment (5 kg/ha).

The experimental findings were as follows;

The emergence of seedling and growth parameters like plant height, no of stems, no of leaves, leaf area was found increase more in the treatment T10 with 75 per cent RD NPK+25 per cent RDN V.C + PSB and Azotobacter than all other treatments including T15, control where it was lowest.

Improvement in available nutrients content i.e, NPK, soil pH, organic carbon, CEC, were seen in most of the treatment from the initial status than in control. The highest Available nutrient content was seen in treatment T1 with 100 per cent vermicompost or treatments which content 75 per cent of vermicompost in it both in 0-15 and15-30 cm depth.

The yield both in grade wise and total yield per plot was recorded highest in treatment T10 with 75 per cent RD NPK+25 per cent RDN V.C +PSB and Azotobacter (19.63 t/ha)

The maximum economic return was associated with the treatment T10, 75 per cent RD NPK+25 per cent RDN V.C + PSB and Azotobacter with maximum net return (193091.5) and B: C ratio (1.9) because of higher value of crop produce under investigation.

Conclusion

On the basis of results from the present investigation, it can be concluded that potato requires adequate nutrition under acid soils of high hill and eastern Indian Himalayan region that may be achieved by judicious and balanced use of organic and inorganic sources of nutrients.

Thus it may be concluded that conjoint use of 75 per cent recommended dose of fertilizers through synthetic fertilizers and remaining 25 per cent recommended dose of fertilizers through vermicompost based on nitrogen basis along with biofertilizers increased the productivity, profitability, reduced the use of chemical fertilizers and also enriched the fertility status of soil for sustainable production of seed tubers.

Thus organic manures and biofertilizers should be a part of the agronomic practices for potato cultivation in Manipur condition.

REFERENCES

1. Burman, R.R., Kumar, M. and Nagraj, K.M. (2007). Organic potato production practices and extension strategies. *In advancing in organic farming technology in india*, pp. 271-279.
2. Meitei, W.I. and Singh, A.I. (2001). Nutrient requirements of important Horticultural Crops of Manipur. In the proceeding of National Seminar on Water and Land Management including CAD for Socio Economic upliftment of North East Region. **2**: 119 -127.
3. Bouyoucos, G.H. (1951). *Soil Scien*ce, 23: 319-314
4. Walkley, A.J. and Black, I.A. (1934). Estimation of soil organic carbon by the chromic acid titration method. Soil Science, 37: 29-38.
5. Jackson, M.L. (1973). Soil Chemical Analysis. Prentice Hall of India Private limited, New Delhi.
6. Subbiah, B.V. and Asija, G.L. (1956). A rapid procedure for estimation of available nitrogen in soils. Current Science, 25: 259-260.
7. Bray, R.H. and Kurtz, L.T. (1945). Determination of total organic and available forms of phosphorus in soils. Soil Science, 59: 39-45.
8. Borah, D.K., Bordoloi, P.K., Karmakar, R.M., Baruah, N.G. and Das, M. (1987). *Practical Manual for Fundamental of Soil Science* (*Part-III*), (*Jorhat, Assam*).

Chapter 10

Influence of Boron and Sulphur on Soil Properties and Yield of Broccoli (*Brassica oleracea* var. *italica*)

Rafiya Choudhary

ABSTRACT

A field experiment was conducted to study effect of boron and sulphur on soil properties and yield of Broccoli (Brassica oleracea var. italica) during rabi season of 2019-20 and 2020-2021 at research farm of Central Agricultural University, Imphal, Manipur. The experiment consists of 12 treatments and two factors. The treatments included three levels of Boron viz., 0, 1 and 1.5 kg ha^{-1}and four levels of sulphur viz., 0, 15, 30 and 40 kg ha^{-1}in combination with RDF. The experiment was laid out following a factorial randomized block design with four replications. The main effect of different level of boron and sulphur was found positive and significant on curd diameter(cm), curd weight(g), days required for curd initiation and curd maturity, nutrient content and yield (t ha^{-1}). The maximum yield (t ha^{-1}), yield parameters and N, P, K, S and S content in plants was obtained from B3 (1.5 kg B ha^{-1}) which was statistically at par with B2 (1 kg B ha^{-1}). Application of sulphur at 30 kg S ha^{-1}obtained maximum yield (t ha^{-1}), yield parameters and N, P, K, S and S content in plants. Maximum net return and B: C ratio was also obtained with the application of 1.5 kg B ha^{-1}and 30 kg S ha^{-1}. However, combined application of 30 kg S ha^{-1} and 1 kg B ha^{-1} (T7) was found superior in terms of yield, yield parameters, nutrient content, net returns (278955 Rs ha^{-1}) and B:C ratio (3.30) compared to that in the control. Hence, it is advisable to the farming community that application of 30 kg S ha^{-1}+ Borax @ 1 kg ha^{-1}was economically profitable for obtaining higher productivity and quality in broccoli besides maintaining Sulphur and boron status in soil.

Correlation studies showed that available boron was positively correlated but found non-significant with pH, EC, OC, N, P and K in 1st and 2nd year. Boron showed positive and significant correlation with Clay and CEC. Correlation studies showed that sulphur was positively and significantly correlated with OC, Clay, N, P and K in 1st and 2nd year. However, sulphur showed positive and non-significant correlation with EC and CEC but found negatively and non-significantly correlated with pH in both years.

Introduction

Broccoli (*Brassica oleracea* var. italica L.) is an important member of "Cole" crops which belongs to the family Brassicaceae. It is also a heavy feeder of plant nutrients. Italy is a centre of diversification (Singh and Nath, 2012). It is a winter season rare vegetable in India, commonly known as Harigobhi and is well distributed to both tropical and sub-tropical regions.There are three types of broccoli, *viz.*,green, white and purple. It is considered to be the most nutritious among all cole crops being rich in vitamin A and mineral besides an anti-cancer compound known as sulphoraphane (Kalia, 1995).

Broccoli is rich source of glucosinolates, the precursor of the chemoprotective isothiocyanate (Sulphoraphane), a compound associated with reducing risk of cancer (Aires *et al.*, 2006). It is a high value exotic vegetable and cultivated for its tender flowering head and the secondary heads *i.e.* spears which develop in the axils of leaves and may contribute up to 50 per cent of the total yield. Generally it can be classified into three distinct group *viz.*, white, purple and green, out of which green type is highly nutritious (Yoldas *et al.*, 2008). It grows best when exposed to an average daily temperature between18-20°C.

India rank second in area and production in Broccoli. World area and production are 1.21 million hectare and 20.88 million tonne and Indian production and area are 6745 thousand tonnes and 369 thousand hectare (Anonymous, 2015). However, broccoli is grown in small scale in Manipur. It is mostly cultivated in hilly areas of H.P., U.P., J and K, Nilgiri hills and Northern plains of India. It is sold in markets as fresh vegetable and used as curries, fried vegetables, soups and pickles. It is fairly high in protein (3.3 per cent), vitamin „A (9000 IU) and ascorbic acid (137 mg/100g). It also contains appreciable quantities of carbohydrate (5.5 per cent), calcium (1.29 per cent), phosphorus (0.79 per cent), sulphur (1.26 per cent), thiamine, riboflavin, niacin and iron (Ramphal, 2000).

Boron (B) is an essential micronutrient which is required for growth of the plant and is directly and indirectly involved in many plant metabolic functions. Boron is an essential component of plants for growth and development. It is a constituents of cell membrane and is essential for cell division. In plants B is required in the structure of cell wall (O Neil *et al.*, 2004).B is the only micronutrient not specifically associated with either photosynthesis or enzyme function. But it is essential for translocation of sugar, starches, nitrogen and sulphur. It helps in proper division and elongation of cell, promotes cell wall strength, flowering, pollination, IAA and carbohydrate metabolism and translocation of sugar and its deficiency may affect all these processes.

Boron deficiency has been realized as the second most important micronutrient constraint in crops after that of zinc (Zn) on global scale (Ahmad *et al.,* 2012).Several soil factors and conditions render soils deficient in boron. It is reported that low soil organic matter content, coarse/sandy texture, high pH, liming, drought, intensive cultivation and more nutrient uptake than application, and the use of fertilizers which are poor in micronutrients are considered to be the major factors associated with the occurrence of boron deficiency (Niaz *et al.,* 2007).

Hollow stem disorder is a major problem for broccoli production and is commonly associated with B deficiency (Shelp *et al.,* 1992). It is well-researched that growth and quality of plants are greatly influenced by a wide range of nutrients. Application of boron significantly increases curd diameter, weight of curd, yield and quality of cauliflower (Kumar *et al.,* 2002).

Sulphur is increasingly being recognized as the fourth major plant nutrient after nitrogen, phosphorus and potassium (Jamal *et al.,* 2010). It is an essential plant nutrient and plays a vital role in biosynthesis of certain amino acids (Cysteine, cystine and methionine) and also helps in the synthesis of co-enzyme-A and formation of chlorophyll and nitrogenase enzyme. It is reported from two natural growth regulators *viz.,* thiamine and biotin. It occurs in glutathione that is important to oxidation reduction reaction.

In recent years, sulphur deficiency has become widespread due to continuous use of sulphur free fertilizers, high yielding varieties and intensive cropping of high sulphur requiring crops. In addition to that leaching, erosion losses also contribute to sulphur deficiency. In India sulphur deficiency has been reported from more than 120 districts which account 30-35 million hectares of crop land (Tandon, 1991).

Kanwar and Mohan (1962) reported that light textured soils are more liable to sulphur deficiency due to their low organic matter content and lesser to protect sulphate from leaching. Hara *et al.* (1981) reported that dry weight was increased with an increase in the sulphur levels in cabbage. They also found that the sulphur deficient plants of cabbage were low in sugars. Jaggi and Dixit (1995) conducted a field experiment on cauliflower and investigated that increasing application of sulphur increased sulphur concentration and uptake in plant

Falk *et al.* (2007) reported that the most striking feature of brassica crops is their high levels of glucosinolates and these molecules contain at least two sulphur atoms, which is the main reason for the high sulphur demand of brassica crops. Mengal and Kirkby (1987) reported that boron requirement is high in cole crops like cauliflower and broccoli.

Due to this reason, the field experiment entitled "Influence of Boron and Sulphur on soil properties and Yield of Broccoli (*Brassica oleracea* var. italica)" was conducted to check the fertility status of soil and find out the optimum dose of sulphur and boron for maximization of yield of broccoli. The objectives are as following:

1. To study the effect of Boron and Sulphur on soil properties and yield of the crop.
2. To study the interaction effect of Boron and Sulphur on soil properties and yield of the crop.
3. To study the correlation of Boron and Sulphur with soil properties.

Materials and Methods

Description of Study Site

The details of the materials and methods of this research work are described in this chapter. The experimental site, climate and weather, experimental design, layout, materials used for experiment, sowing of seeds, intercultural operations, harvesting, collection of data and statistical analyses are given below:

Experimental Site

The experiment was conducted at the research field in College of Agriculture, Central Agricultural University, Iroishemba, Imphal, Manipur during the Rabi season of 2019-2020 and 2020-2021. The climatic condition of Imphal is Sub-tropical.

Soil of the Experimental Field

Composite soil samples were collected at random from the depth of 0- 15cm from the experimental field before starting the experiment. A homogenous composite sample was then prepared for mechanical and chemical analysis. Result of the analysis and method used for determination of the soil properties are presented in Table 10.1.

Table 10.1. Physico-chemical Parameters of the Initial Soil of the Experiment

Parameters	*Value*	*Status*	*Method Followed*
A. Physical properties			
1. Soil texture			Hydrometer method (Bouyoucos, 1962)
i) Sand (per cent)	17.5		
ii) Silt (per cent)	31.8	Clay	
iii) Clay (per cent)	50.7		
Chemical properties			
1. pH	5.6	Acidic	1:2.5 soil: water suspension (Jackson, 1973)
2. CEC	11.12	Medium	Conductometery (Jackson, 1973)
3. EC	0.23	Low	Conductometery (Jackson, 1973)
4. Organic carbon (per cent)	1.89	High	Walkley and Black titration (Walkley and Black, 1934)

Parameters	*Value*	*Status*	*Method Followed*
5. Available N (kg/ha)	313.21	Medium	Alkaline Permanganate (Subbiah and Asija,1956)
6. Available P_2O_5 (kg/ha)	26.12	Low	Bray s method (Bray and Kurtz, 1945)
7. Available K_2O (kg/ha)	236.52	Medium	Flame photometer (Jackson, 1973)
8. Available Sulphur (mg/kg)		High	Turbidimetric method (Williams and Steinbergs, 1959)
9.Available Boron (mg kg^{-1})	0.23	Low	Hot water method (Berger and Troug, 1939)

Experimental Details

Treatment Details

The experiment was undertaken to study the effect of 3 levels of boron and 4 levels of sulphur on the growth and yield of broccoli. The study comprises the following treatment.

Boron Level: 3	*Sulphur level: 4*
1) B_1:0(Control)	1) S_1: 0(Control)
2) B_2: 1kg B ha^{-1}	2) S_2: 15 Kg ha^{-1}
3) B_3: 1.5kg B ha^{-1}	3) S_3: 30Kg ha^{-1}
	4) S_4: 40 kg ha^{-1}

There were 12 treatment combinations of boron and sulphur levels used in the experiment as followings:

B_1S_1	T_1	Control
B_1S_2	T_2	0 kg B/ha + 15 kg S ha^{-1}
B_1S_3	T_3	0 kg B/ha + 30 kg S ha^{-1}
B_1S_4	T_4	0 kg B/ha + 40 kg S ha^{-1}
B_2S_1	T_5	1 kg B/ha + 0 kg S ha^{-1}
B_2S_2	T_6	1 kg B/ha + 15 kg S ha^{-1}
B_2S_3	T_7	1 kg B/ha + 30 kg S ha^{-1}
B_2S_4	T_8	1 kg B/ha + 40 kg S ha^{-1}
B_3S_1	T_9	1.5 kg B/ha + 0 kg S ha^{-1}
B_3S_2	T10	1.5 kg B/ha + 15 kg S ha^{-1}
B_3S_3	T11	1.5 kg B/ha + 30kg S ha^{-1}
B_3S_4	T12	1.5 kg B/ha + 40 kg S ha^{-1}

Design and Layout of the Experiment

The experiment was laid out in asymmetrical factorial randomized block design with four replications. The experimental plot was first divided into four blocks. Each block consisted of 12 units of plots. The size of each plot was 8.4 m^2. The total no. of plots were 48. Different combination of S and B were assigned randomly to each block as per design of the experiment.

Crop	Broccoli
Variety	Green Magic
Number of treatments	12
Number of replication DesignTotal number of plots	4Factorial randomised block design 48
plots size	3 m x 2.8 m(8.4m^2)
Spacing	60 cm X 45 cm
Borders between the replication	0.3 cm
Borders between the plots	0.5 cm

Cultivar Description

The variety selected for experiment was "Green Magic" which is of mid to early maturity class. It is harvested in about 90 days from sowing. It is semi-dome, tight head with medium small bead in bluish green color. It has good heat tolerance and average cold tolerance. It is also tolerant to bacterial leaf spot, black rot, downy mildew and hollow stem.

Detail of Crop Rising

Nursery Bed Preparation

Seed bed of 2 x 1 m (length x breadth) was prepared and seed was sown in 5 cm line spacing and 2 cm as a depth. Seed was sown on 4th of October 2019 for 1st year and 10th of October 2020 for 2nd year and covered with loose soil.

Land Preparation

Area of experiment was ploughed deeply with tractor for 2 times followed by power tiller and rotavator for 1 time before transplanting, to bring the soil fine tilth. The weeds and stubbles were cleaned by rache and finally and was levelled. Then, the field was divided into plots as per required dimensions by using the measuring tape, rope and pegs. A distance of 0.5 m kept between two plots.

Table 10.2. Physico-chemical Parameters of the Initial Soil of the Experiment

Parameters	*Value*	*Status*	*Method Followed*
A. Physical properties			
1. Soil texture			Hydrometer method (Bouyoucos, 1962)
i) Sand (per cent)	17.5	Clay	
ii) Silt (per cent)	31.8		
iii) Clay (per cent)	50.7		
B. Chemical properties			
1. pH	5.6	Acidic	1:2.5 soil: water suspension (Jackson, 1973)
2.CEC	11.12	Medium	Conductometery (Jackson, 1973)
3. EC	0.23	Low	Conductometery (Jackson, 1973)
4. Organic carbon (per cent)	1.89	High	Walkley and Black titration (Walkley and Black, 1934)
5. Available N (kg/ha)	313.21	Medium	Alkaline Permanganate (Subbiah and Asija, 1956)
6. Available P_2O_5 (kg/ha)	26.12	Low	Bray s method (Bray and Kurtz, 1945)
7. Available K_2O (kg/ha)	236.52	Medium	Flame photometer (Jackson, 1973)
8. Available Sulphur (mg/kg)		High	Turbidimetric method (Williams and Steinbergs, 1959)
9.Available Boron (mg kg^{-1})	0.23	Low	Hot water method (Berger and troug, 1939)

Experimental Details

Treatment Details

The experiment was undertaken to study the effect of 3 levels of boron and 4 levels of sulphur on the growth and yield of broccoli. The study comprises the following treatment.

Boron Level: 3	*Sulphur Level: 4*
1) B_1:0(Control)	1) S_1: 0(Control)
2) B_2: 1kg B ha^{-1}	2) S_2: 15 Kg ha^{-1}
3) B_3: 1.5kg B ha^{-1}	3) S_3: 30Kg ha^{-1}
	4) S_4: 40 kg ha^{-1}

There were 12 treatment combinations of boron and sulphur levels used in the experiment as followings:

B_1S_1	T_1	Control
B_1S_2	T_2	0 kg B/ha + 15 kg S ha^{-1}
B_1S_3	T_3	0 kg B/ha + 30 kg S ha^{-1}
B_1S_4	T_4	0 kg B/ha + 40 kg S ha^{-1}
B_2S_1	T_5	1 kg B/ha + 0 kg S ha^{-1}
B_2S_2	T_6	1 kg B/ha + 15 kg S ha^{-1}
B_2S_3	T_7	1 kg B/ha + 30 kg S ha^{-1}
B_2S_4	T_8	1 kg B/ha + 40 kg S ha^{-1}
B_3S_1	T_9	1.5 kg B/ha + 0 kg S ha^{-1}
B_3S_2	T10	1.5 kg B/ha + 15 kg S ha^{-1}
B_3S_3	T11	1.5 kg B/ha + 30kg S ha^{-1}
B_3S_4	T12	1.5 kg B/ha + 40 kg S ha^{-1}

Design and Layout of the Experiment

The experiment was laid out in asymmetrical factorial randomized block design with four replications. The experimental plot was first divided into four blocks. Each block consisted of 12 units of plots. The size of each plot was 8.4 m^2. The total no. of plots were 48. Different combination of S and B were assigned randomly to each block as per design of the experiment.

Crop	Broccoli
Variety	Green Magic
Number of treatments	12
Number of replication DesignTotal number of plots	4Factorial randomised block design 48
plots size	3 m x 2.8 m(8.4m^2)
Spacing	60 cm X 45 cm
Borders between the replication	0.3 cm
Borders between the plots	0.5 cm

Cultivar Description

The variety selected for experiment was "Green Magic" which is of mid to early maturity class. It is harvested in about 90 days from sowing. It is semi-dome, tight head with medium small bead in bluish green color. It has good heat tolerance and average cold tolerance. It is also tolerant to bacterial leaf spot, black rot, downy mildew and hollow stem.

Detail of Crop Rising

Nursery Bed Preparation

Seed bed of 2 x 1 m (length x breadth) was prepared and seed was sown in 5 cm line spacing and 2 cm as a depth. Seed was sown on 4th of October 2019 for 1st year and 10th of October 2020 for 2nd year and covered with loose soil.

Land Preparation

Area of experiment was ploughed deeply with tractor for 2 times followed by power tiller and rotavator for 1 time before transplanting, to bring the soil fine tilth. The weeds and stubbles were cleaned by rache and finally and was levelled. Then, the field was divided into plots as per required dimensions by using the measuring tape, rope and pegs. A distance of 0.5 m kept between two plots.

Transplanting

Twenty five days old, uniform and healthy seedlings were selected and transplanted in the experimental plots. The seedbed were watered one hour before uprooting the seedlings to avoid damage to the root system. Transplanting was done in the afternoon to escape the transplanted seedlings from shock. Seedlings were transplanted at a distance of 60 cm x 45 cm. Thus, each unit plot accommodated 25 plants. The transplanted seedlings received watering for 7 days for their proper establishment in the field.

Fertilizer Applications

The recommended doses of nitrogen, phosphorus and potash were applied during final land preparation in the form of urea, diammonium phosphate and muriate of potash respectively. During the calculation of amount of urea and diammonium phosphate per plot, nitrogen and phosphorous supplied through diammonium phosphate and Single Super Phosphate respectively, were subtracted. Plots were irrigated immediately after the application of fertilizers. Sulphur was applied 21 days before transplanting.

Nutrients	*Dose/ha*	*Source*
Boron (B)	0, 1, 1.5 Kg B/ha	Borax (11 per cent B)
Sulphur(S)	0, 15, 30, 40 Kg S/ha	SSP (12 per cent S)
Nitrogen (N)	120 Kg N/ha	Urea (46 per cent N)
Phosphorus(P)	80 Kg P_2O_5/ha	DAP (18 per cent N,46 per cent P)
Potassium (K)	40 Kg K_2O/ha	MOP (60 per cent K)

Intercultural Operations

Intercultural operations were performed as and when necessary throughout the growth period of the crop.

Gap Filling

The transplanted seedlings were kept under observation and gap filling was done by the seedlings from the same nursery to maintain desired plant density.

Irrigation

First irrigation was given immediately after transplanting. Irrigation was given as per requirement of the crop in order to maintain the soil moisture because Broccoli requires continuous supply of moisture. Heavy irrigation is avoided at the time of irrigating the crop.

Earthing Up

Earthing up at 35 and 50 days after transplanting (DAT) around the plants was done by taking the soil from the space between the rows by a small spade.

Weeding and Hoeing

The crop kept weed free from weeds throughout its growth period to avoid weed competition, for better aeration and to break the soil crust. It helped to increase soil moisture conservation. Broccoli roots are surface feeders; hence care was taken at the time of weeding.

Harvesting

The mature curds were harvested with 2-4 cm stalk by a sharp knife and weight taken on electronic weighing balance. Before harvesting of the broccoli head, compactness of the head was judged by pressing with thumbs. Heads were labelled and the observations were recorded.

Biometric Observation

Five plants were randomly selected from the middle rows of the unit plots for avoiding border effect, except yield of curd, which were recorded plotwise. Selected plants were tagged from each treatment under each replication for recording the observations for the following parameters.

Days Required for Curd Initiation after Transplanting

Total number of days from the date of transplanting to the date of visible curd initiation was recorded by counting the number of days required by five observational plants in each net plot from transplanting till the initiation of curd. Average numbers of days to first initiation was calculated.

Days Required for Curd Maturity

Days required for curd maturity by five observational plants in each net plot was recorded by counting the total number of days from transplanting to curd maturity stage. Average numbers of days was calculated.

Curd Diameter (cm)

The circumference of curd head at harvest was measured in cm with a thread and later diameter of curd was calculated by using the following formula and the diameter of curd was worked out.

where,

$C = 2 \pi r$

$Cd = \pi$

C = Circumference of the curd (cm) r = Radius of curd (cm)

$\pi = 3.14159$

d = diameter of the curd (cm)

After the full development of the curd and when the clusters were still compact, it was detached from the plants with the help of sharp knife. At each harvest, all the curds harvested from the net plot of observational plants were counted and their diameter was measured with the help of measuring scale and after computing its mean and it was recorded as curd diameter in centimetre.

Curd Weight (g plant^{-1})

After harvesting curd weight was recorded from all plots by removing curd with 2-3 upper leaves from the whole plant and it was recorded using an electronic balance in gm.

Curd Yield (t ha^{-1})

After harvesting from each of the plots curd were removed from the plant with 2-3 upper leave. The weight of curd recorded and calculated to tons per hectare basis.

Curd yield (t ha^{-1}) = Plot yield (kg) × 10000 Plot size (m^2) × 1000

Soil Analysis after Harvest

Soil samples were collected at 0-15 cm soil depths from five locations of each plot of the experimental field before and after the harvesting of crop. A homogenous composite sample was then prepared for the mechanical and chemical analysis.

Soil Texture

Soil texture of the soil was done by hydrometer method (Bouyoucos,1962). The textural class was determined using Marshall s Triangular Coordinate as designated by USDA (1951).

Soil pH

The glass electrode systronic pH meter was used to determine the pH of the soil samples of each plot. 1:2.5 Soil: Water suspension method as described by Jackson(1973).

Soil Electrical Conductivity

The soil electrical conductivity (EC) was determined by systronic direct reading conductivity meter in 1:2.5 soil: water suspension method as described by Jackson (1973).

C.E.C

Cation exchange capacity of soil samples of each plot was estimated by ammonium saturation methods (Borah *et al.,* 1987).

Organic Carbon (Per cent)

Organic carbon in soil was determined by Walkley and Black s method (Walkley and Black, 1934). The underlying principle is to oxidize the organic carbon present in organic matter with potassium dichromate in presence of sulphuric acid and to titrate the residual $K_2Cr_2O_7$ solution with standard ferrous ammonium sulphate. The result was expressed in percentage.

Available Nitrogen (kg ha^{-1})

Available nitrogen in soil of each plot was determined by alkaline permanganate method (Subbiah and Asija, 1956) in which the soil is mixed with excess of alkaline $KMnO_4$ solution and distilled. Organic matter present in the soil is oxidised by nascent oxygen liberated by $KMnO_4$ in the presence of NaOH and thus ammonia released is distilled and is absorbed in a known volume of standard acid, excess of which is back titrated with standard alkali using methyl red indicator.

Available Phosphorus (kg ha^{-1})

Available phosphorus was estimated by Bray s and Kurtz No.1 method (Bray and Kurtz, 1945).

Available Potassium (kg ha^{-1})

The estimation of available potassium content in the soil samples were done by adopting flame photometer method as described by Jackson (1973)

Available Sulphur (mg kg^{-1})

Available sulphur in soil was determined by turbidimetric method as described by Williams and Steinbergs (1959).

Available Boron (mg Kg^{-1})

Available B status in soil sample was estimated by hot water method as described by Berger and Trough (1939).

Plant Sample Analysis

Plant samples from fully developed curd were harvested, decontaminated, washed first in running tap water, followed by dilute hydrochloric acid (HCl) solution and finally with deionized water and washed with distilled water. The collected plant materials were dried first in air and then in oven at 60°C for 48hrs and ground to a 20 mesh size.Samples were then ground with the help of a stainless steel mechanical grinder.

Determination of Plant Nitrogen

For determination of plant nitrogen, 0.5 g of the plant sample was transferred to the digestion tube. 5g of the catalyst mixture (K_2SO_4, $FeSO_4$, $CuSO_4$) and 10ml of concentrated sulphuric were added in it and the contents of the tube were digested until the organic matter was destroyed and solution had light green colour or colourless. In a 250ml conical flask containing 20ml of 4 per cent boric acid with 2-3 drops of mixed indicator (0.5 per cent bromocerol green and 0.1 per cent methyl red) was placed under the delivery tube of the condenser. Then, the distillation tube containing digested samples was connected to the mouth of distillation flask and 40ml of 40 per cent NaOH was attached to the distillation apparatus and distillation started. Nearly 150ml of distillate is collected in about 8 minutes in the conical flask were titrated against 0.02N H_2SO_4. The end point was the change of colour from green to pinkish colour.

The nitrogen content in plant sample is calculated at,

$$\text{Nitrogen (\%)} = \frac{\text{Titration value X Normality of } H_2SO_4 \text{ X Volume of digestion X100}}{\text{Weight of sample (g) X Aliquot (ml) taken}}$$

Determination of Plant Phosphorus

Phosphorus content in the digested plant was determined by Vanado- molybdo-phosphoric acid yellow colour method using spectrophotometer at 470 nm wave length (Jackson 1973).

The phosphorous content in plants samples is calculated at,

$$\text{Phosphorous (\%)} = \frac{\text{C X Volume of digestion (ml) X 100}}{\text{Weight of sample (g) X Aliquot (ml) taken}}$$

C = Concentration of phosphorous in aliquot (obtained from standard curve) = C ppm

Determination of Plant Potassium

Potassium content in broccoli curd was digested with di-acid, HNO_3:$HClO_4$ and then potassium present in the plant was determined by flame photometry (Jackson 1973).

Determination of Plant Boron

Plant boron was estimated by the dry ashing method which was done in muffle furnace at 550° C for 5 hrs followed by diluting sample with 0.36N H_2SO_4 for 1 hrs. at room temperature as described by Gain and Mitchel,1979 and subsequent filtration by whatman no.1 by using Azomethine-H method (Wolf, 1971).

Determination of Plant Sulphur

Sulphur is estimated in plants digest obtained by wet digestion with HNO_3 and $HClO_4$.The procedure adopted for estimation of sulphur content in plant was preparation of standard curves same as for sulphur estimation in soil. 1g of plant sample is digested in 10-15 ml of di-acid and make up the volume to 100ml. Then, transferred 10ml of the aliquot to a 100ml volumetric flask and add 1g of sieved $BaCl_2$, shake it for 1 minute and add 1ml of gum acacia acetic acid solution, make up the volume and again shake it for 1minute. A blank sample was also run in an identical manner. After 25-30 minutes of precipitation at 440nm the turbidity was measured. Then the sulphur content in the sample was read from the standard curve against the similar absorbance as noted for the sample.

Statistical Analysis of Data

The experimental data obtained were subjected to statistical analysis by adopting Fisher s methods of analysis of variance. The level of significance used in „F test was given at 5 percent of probability. Critical difference (CD) values are given in the table at 5 per cent level of significance whenever the „F test was significance.

Pooled analysis of variance was done for the experimental data after testing the homogeneity of variance by F- Maximum (Gomez and Gomez, 1986).

The experiment was laid out in factorial randomized block design (FRBD) with four replications of each treatment. In all there were sixteen treatment combinations. The unit plot size of the field was 8.4 m^2(3m x 2.8m).The treatments included three levels of boron *viz.,* B_1 (0 kg B ha^{-1}), B_2 (1 kg B ha^{-1}) and B_3 (1.5 kg B ha^{-1}) and four level of sulphur *viz.,* S_1 (0 kg S ha^{-1}), S_2 (15 kg S ha^{-1}), S_3 (30 kg Sha^{-1}) and S_4 (40 kg S ha^{-1}. Five plants will be selected at random in such a way that the border effect could be avoided. Data on yield and yield attributes were recorded from the selected plants. Based on the findings of the experiment the individual and interaction effect of boron (B) and sulphur(S) on soil properties and crop yield were significantly influenced. The findings of this experiment are summarized as under.

Effect of Boron

Effect of boron on yield and yield attributes and nutrient content was found significantly increase from control but the two level boron was found non-significant.

- ✰ Application of boron level B_3 (1.5Kg ha^{-1}) obtained maximum curd diameter in both years which is at par with the boron level B_2 (1kg ha^{-1}) whereas lowest curd diameter was obtained from the boron level B_1 (control).

- Maximum curd weight was obtained from the plants which was treated with the boron level B_3 (1.5Kg/ha) in both year while minimum curd weight was obtained from the application of boron level B_1(control).
- Application of boron level B_3 (1.5Kg ha^{-1}) obtained highest yield from both years which is at par with the boron level B_2 (1 kg ha) whereas lowest yield was obtained from the boron level B_1 (control).
- Minimum days taken to curd initiation was found in both years from the application of boron level B_3 (1.5Kg ha^{-1}) while more number of days taken for curd initiation was obtained from the boron level B_1 (control).
- Days taken for curd maturity was found maximum from both years with the application of boron level B_3(1.5 kg ha^{-1}) whereas minimum number of days taken was obtained from the plants receiving boron level B_1(control).
- The highest nutrient content (N, P, K, S and B) in plants was found in both years with the application of boron level B_3 (1.5 kg ha^{-1}) which is found non-significant with the application of boron level B_2 (1kg ha^{-1}). However, the lowest nutrient content was found from the boron level B_1 (control).Application of boron level B_3 (1.5 kg ha^{-1}) fetched maximum net return and B: C ratio Thus, application of boron level B_3(1.5kg B ha^{-1}) is advisable for broccoli crop.

Effect of Sulphur

Effect of sulphur on yield, yield attributes and nutrient content was found significant

- Maximum curd diameter was obtained with the application of sulphur level S_3 (30 kg ha^{-1}) in both years whereas lowest curd diameter was obtained from the sulphur level S_1 (control).
- Maximum curd weight was obtained from the plant which was treated with the sulphur level S_3 (30 kg ha^{-1}) in both years while minimum curd weight was obtained from the application of sulphur level S_1 (control).
- Application of sulphur level S_3 (30 kg ha^{-1}) obtained highest yield from both years whereas lowest yield was obtained from the sulphur level S_1 (control).
- Application of sulphur level S_3 (30 kg ha^{-1}) reduce the days taken to curd initiation in both years while minimum days taken for curd initiation increases with the application of the sulphur level S_1 (control).
- Days taken for curd maturity was found maximum from both years with the application of sulphur level S_3 (30 kg ha^{-1}) whereas minimum number of days taken was obtained from the plants receiving sulphur level S_1 (control).

- The highest nutrient content (e.i. N, P, K, S and B) in plants was found in both years with the application of sulphur level S3(30 kg ha^{-1}). However, the lowest nutrient content was found from the sulphur level S1 (control).
- Application of sulphur level S3(30 kg ha^{-1}) fetched maximum net return and B:C ratio compared to other sulphur level and control.

Interaction Effect of Boron and Sulphur

- Interaction effect of boron and sulphur on crop yield and yield attributes was found significantly increase with increase in sulphur and boron. However, the interaction becomes antagonistic at highest level of sulphur 40 kg ha^{-1} with 1 kg and kg B ha^{-1}.
- The treatment combination of B3S2 Kg ha^{-1}(T7) gave maximum curd diameter whereas the minimum curd diameter was obtained from the control treatment (T1).
- Maximum curd weight was recorded from the treatment combination of B3S2 Kgha^{-1} (T7) while minimum curd weight was obtained from the plant receiving no boron and sulphur (T1).
- The treatment combination of S3B2 (T7) gave maximum yield per ha whereas minimum yield was obtained from the control treatment (T1).
- Minimum days taken for curd initiation was found from the treatment combination of S3B2 (T7) and maximum days taken for curd initiation was found from the control treatment (T1).
- A maximum day taken for curd maturity was obtained from the treatment combination of S3B2 (T7) whereas a minimum day taken for curd maturity was found from the control treatment (T1).
- Highest nitrogen, phosphorous, potassium, sulphur and boron content in plants was obtained from the treatment combination S3B2 (T7) while minimum nitrogen, phosphorous, potassium, sulphur and boron content was found from the control treatment (T1).
- Combined application of 1 kg B ha^{-1} and 30 kg S ha^{-1} fetched highest net return and B: C ratio.

Boron and Sulphur Correlation with Soil Properties

- Correlation studies revealed that boron was positively correlated but statistically non-significant with soil pH (r=0.607), EC (r=0.443), OC(r=0.464) in 1st year and pH (r=0.624), EC (r=0.485), OC (r =446) in 2nd year. However, CEC (r=0.947*) and clay (r=0.951*) in 1st year and CEC (r=0.951*) and clay (r=0.947*) in 2nd year show positive and significant correlation with soil available boron. Nitrogen (r=0.514), phosphorous (r=0.515) and potassium (r=0.227) in 1st year and nitrogen (r=0.515), phosphorous(r=0.527) and potassium (r=0.447) in 2nd year show positive but non- significant correlation with available boron.

- ☆ Correlation studies showed that sulphur was negatively correlated but statistically non-significant with soil pH (r=-0.486) in 1st year and (r=-180) in 2nd year. CEC (r=0.558) in 1st year and (r=0.574) in 2nd year is positively but non-significantly correlated. However, EC (r=0.913*), clay (r=0.902*), OC (r=0.908*), nitrogen (r=0.902*), phosphorous (r=0.906*) and potassium (r=0.905*) were positively and significantly correlated in 1st year and EC (r=0.928)*), clay (r=0.910*), OC (r=0.913*) nitrogen (r=0.906*), phosphorous (r=0.910*) and potassium (r=0.907*) in 2nd year of soil.

Conclusion

- ☆ From the overall results of the present investigation, it may be concluded that the individual application of boron level B3 (1.5 kg B ha^{-1}) was found best in terms of yield, yield attributes which was found statistically at par with the application of boron level B2 (1 kg B ha^{-1}).Highest net return and B:C ratio was obtained with application of boron level B3(1.5 kg B ha^{-1}). Individual effect of sulphur was also found significant in terms of yield, yield attributes, net return and B: C ratio. Highest yield was obtained with the application of sulphur level S3 (30 kg S ha^{-1}). Combined application of boron and sulphur was found significant. Highest yield, net return and B: C ratio was obtained from the treatment combination of S3B2 (T7). In case of interaction, it is advisable to farming community that 1 kg B ha^{-1} and 30 kg S ha^{-1} was economically profitable for obtaining optimum yield of broccoli.

REFERENCES

1. Singh, D.N. and Nath, V. (2012). Winter Vegetables: Advances and Developments. Satish Publishing House, Delhi (India), p. 360.
2. Kalia P (1995).Biochemical assays and yield performance of sprouting broccoli genotypes of Himachal Himalayas. *Cruciferae Newsl.*, 17: 90-91.
3. Aires, A., Rosa, E., and Carvalho, R.(2006). Effect of nitrogen and sulfur fertilization on glucosinolates in the leaves and roots of broccoli sprouts (*Brassica oleracea* var. *italica*). *J. Sci. Food Agri.*, 86(10): 1512-1516.
4. Yoldas, F., Ceylan, S., Yagmur, B., and Mordogan, N. (2008). Effects of nitrogen fertilizer on yield quality and nutrient content in broccoli. *J. Pl. Nut.*, 31(7): 1333-1343.
5. Anonymous. (2015) Indian Horticulture Database, National Horticulture board. Govt. of India Ministry of agriculture, 144-151.
6. Ramphal (2000) Gobhi vargiya sabgiyan. ICAR, New Delhi. Pp Jordon, H.V. and Reisenaur,

7. H.M. (1957). Sulphur and soil fertility. Soil year book. USDA, Washington. pp 107-111.

8. O'Neill, M.A., Ishii, T., Albersheim, P., and Darvill, A.G. (2004). Rhamnogalacturonan II: structure and function of a borate cross-linked cell wall pectic polysaccharide. *Annu. Rev. Plant Biol.,* 55: 109-139.

9. Ahmad, W., Zia, M.H., Malhi, S.S., Niaz, A. and Ullah, S. (2012). Boron deficiency in soils and crops: a review. *Crop Pl.,* pp 65-97.

10. Niaz, A., Ranjha, A., Rahmatullah, M., Hannan, A. and Waqas, M. (2007). Boron status of soils as affected by different soil characteristics–pH, CaCO3, organic matter and clay contents. *Pak. J. Agric. Sci.,* 44(3): 428-435).

11. Shelp, B.G., Shattuck, V.I., Lettan, D. Mc. and Liu, L. (1992). Boron nutrition and the composition of glucosinolate and soluble nitrogen compound in two broccoli cultiver. *Can. J. Plant Sci.,* 72: 889-89.

12. Kumar, S., Chaudhury, D.R. and Kumar, S. (2002). Effect of FYM, molybdenum and boron application on yield attributes and yield of cauliflower. *Crop Research-Hisar,* 24(3): 494-496.

13. Jamal, A., Moon, Y.S., and Abdin, M.Z. (2010). Sulphur-A general overview and interaction with nitrogen. Australian. *J. Crop Sci.,* 4(7): 523-529.

14. Tandon, H.L.S. (1991). Sulphur research and agriculture production in Indian. Fertilizer Development and Consultation Organization C-110, Greater Kailash-1, New Delhi.

15. Kanwar, J.S. and Mohan (1962). Distribution of forms of sulphur in Punjab soils. symposium on Fertility of Indian Soils, 26: 37-38.

16. Hara, J., Sugimoto, K. and Sonoda, Y. (1981). Nutritional relationship between nitrogen and sulphur in cabbage (*Brassica oleracea* var. *capitata*). *J. Japan Soc. Hortic. Sci.,* 50(*1*):60-65.

17. Falk, K.L.; Tokuhisa, J.G. and Gershenzon, J. (2007). The effect of sulfur nutrition on plant glucosinolate content: Physiology and molecular mechanisms. *Plant Biol.,* 9: 573–581.

18. Bouyoucos, G.H. (1962). The hydrometer as a new method for the mechanical analysis of soils. *J. Agron.,* 54: 464-465.

19. Jackson, M.L. (1973). Soil Chemical Analysis. Prentice Hall of India Private limited, New Delhi.

20. Walkley, A.J. and Black, I.A. (1934). Estimation of soil organic carbon by the chromic acid titration method. Soil Science, 37: 29-38.

21. Subbiah, B.V. and Asija, G.L. (1956). A rapid procedure for estimation of available nitrogen in soils. Current Science, 25: 259-260.

22. Bray, R.H. and Kurtz, L.T. (1945). Determination of total organic and available forms of phosphorus in soils. Soil Science, 59: 39-45.

23. Williams, C.H., and Steinbergs, A. (1959). Soil sulphur fractions as chemical indices of available sulphur in some Australian soils. *Aus. J. Agric. Res.*, 10(3): 340-352.

24. Berger, K.C., and Truog, E. (1944). Boron tests and determination for soils and plants. *Soil Sci.*, 57: 25-36.

25. Borah, D.K., Bordoloi, P.K., Karmakar, R.M., Baruah, N.G. and Das, M. (1987). *Practical Manual for Fundamental of Soil Science* (*Part-III*), (*Jorhat, Assam*).

Index

9 789359 191348

Printed at Repro India Ltd.